RISE OF THE MINING ROYALTY COMPANIES

ROYALTIES STREAMS

2004 2005 2006 2007 2008 2009 2010 2011 2012 2013 2014 2015 2016 2017 2018 2019 2020 2021

Edited by Douglas Silver

PUBLISHED BY THE
SOCIETY FOR MINING, METALLURGY & EXPLORATION

Society for Mining, Metallurgy & Exploration (SME)
12999 E. Adam Aircraft Circle
Englewood, Colorado, USA 80112
(303) 948-4200 / (800) 763-3132
www.smenet.org

SME advances the worldwide mining and minerals community through information exchange and professional development. SME is the world's largest association of mining and minerals professionals.

ISBN 978-0-87335-497-4
Ebook 978-0-87335-498-1

Library of Congress Cataloging-in-Publication Data has been applied for.

CONTENTS

Urban myth: The word "royalty" is a contraction of "royal tithing." The miner had to pay the king a small tithing to mine on the king's lands.

PREFACE

Royalties have been imposed since Roman times (400 AD) when a 10% tax was levied on privately owned mines. The English and French were the first European states to codify the practice around 1300 AD. Today, the creation of royalties is most prevalent in British Commonwealth or former colonies, although the governments of a number of countries around the world have adopted royalties as part of the payments owed to them by mine owners when mining projects go into production. Private royalties have been created by contracts in many jurisdictions and often have the same characteristics as government royalties.[1]

The modern public mining royalty and streaming industry began in 1986 with the conversion of Franco-Nevada Mining Corporation from an exploration into a royalty company. Collectively, this industry has since grown to more than 35 companies and approximately $66 billion of market capitalization.[2]

The royalty and streaming industry offers two important services. It is a major funding source to large and small companies, and its public companies command some of the higher price-to-net-asset-value (P/NAV) ratios of any industry.

This book is not intended to be a primer on the basics of royalties and streams (although a short overview is provided). It is assumed that the reader has a working knowledge of these instruments. The book's focus is to show readers what the founders were thinking about the industry at the time of their corporate formation, and how the industry has evolved over the past 36 years.

---

1. International Royalty Corporation, *Annual Information Form*, March 23, 2009, p. 6.
2. These numbers are as of December 31, 2021.

For simplicity's sake, the public mining royalty and streaming industry and its participants are referred to as the "royalty companies" and "royalty industry," respectively. This book only studies mineral offtakes, royalties, and streams owned by public mineral companies. The company chapters are presented in the chronological order of when they became royalty companies. All currencies are presented in United States dollars, unless otherwise noted.

Most of the founders of the largest companies are now retired, and this book captures their stories for future generations. Their intimate knowledge and leadership provide current and future royalty holders and investors with unique knowledge about how this industry has developed, evolved, and thrived.

ACKNOWLEDGMENTS

This book is a collaborative effort by the leaders of the public mining royalty and streaming industry. They have taken their precious time to contribute important information about how they built their companies, adapted to market changes, and continually innovate to advance the industry.

A big "thank you" to all of the people who contributed to this seminal work. This includes, but is not limited to (in order of appearance): Pierre Lassonde, David Harquail, Paul Brink, Donna Andrejek, Lloyd Hong, Stan Dempsey Sr., Tony Jensen, Bill Heissenbuttel, Yelena Ovanesyan, Lori McCasey, Brian Dalton, Lawrence Winter, Randy Smallwood, Simona Antolak, Patrick Drouin, Doug Hurst, Nolan Watson, Kim Bergen, Mark Klausen, Sean Roosen, Sandeep Singh, Heather Taylor, Andre Le Bel, Brett Heath, Drew Clark, Sunny Sara, Ross Beatty, Geoff Burns, Dan O'Flaherty, Brent Bonney, Amir Adani, David Garofalo, Peter Behncke, Shaun Usmar, Steve Bristo, James Dendle, Simon Watkins, Rachel Dare, Brendan Yurik, Kim Lim, Frederick Bell, Alexandra Woodyer Sherron, Kaitlin Taylor, Dave Cole, Scott Close, Teo Dechev, Anthony Milewski, Alex Tsukernik, Akiba Leisman, Peter Bures, Adam Davidson, Scott Melbye, Kyle Floyd, Spencer Cole, and especially Jane Olivier, Terese Platten, Karen Ehrmann, David Hammond, Matt Dugaro and the CIBC team, and Rick Winters.

I would also like to thank SME for publishing this work.

1

OVERVIEW OF THE INDUSTRY

Douglas Silver

The contemporary mining royalty industry effectively began in 1986 when Franco-Nevada Mining Corporation Limited, with a $25 million market capitalization, converted itself from an exploration into a royalty company. Since then, dozens of companies have been created. Currently, there are 36 active public mining royalty companies, representing more than $66 billion in market value, along with an unknown number of private companies.

Just because a company owns a royalty or metal stream doesn't make it a royalty company. Royalty companies are defined as corporations that deliberately and actively create and/or acquire royalties and streams, and the majority of their income is derived from these assets. In most cases, they are easy to identify them because of their activities or because the investment banks track them as royalty companies. In borderline cases, their company name and their corporate description help separate them from the non-royalty companies.

However, because of the high-trading market multiple received by royalty companies, there are players who pretend to be royalty companies but are actually just exploration companies that hold a few royalties. These companies have been excluded from this discussion.

ASSET CHARACTERISTICS AND DEMOGRAPHICS

The true royalty companies that exist today own three types of assets: royalties, streams, and offtakes. These form a continuum of financial instruments with a singular commonality: They extract revenues from a portion of the mine's production and do so as passive investors. This section reviews the characteristics and demographics for assets owned by the royalty companies.

A royalty company can acquire these assets in two different ways. They can work directly with the mine owner and create these instruments as a financing vehicle, or they can purchase pre-existing instruments in the marketplace. These market purchases are most commonly done in portfolio acquisitions but can also be made for single assets.

Royalties

Royalties are created by two principal mechanisms. A prospector discovers a mineral deposit but has neither the technical skill, inclination, nor financial ability to build and operate the mine, so the property is sold to a mining company. The prospector keeps a small stake of the mine's future economics in the form of a royalty.

A second method is more popular today. Mineral companies regularly sell properties that do not meet their corporate objectives but retain a royalty just in case they are proved wrong. They then periodically bundle these contingent assets up and sell them to one of the royalty companies. Royalty companies employ very low discount rates in their valuation models, so the seller often receives a price superior to their internal evaluation.

The four general types of royalties, each with many variations, are described next.

Gross Revenue

A gross revenue royalty (GRR), also known as a *gross overriding royalty, gross metal royalty, gross sales royalty, gross product royalty, gross value royalty, net sales royalty* (including FOB royalties), *overriding royalty*, or a *sales return royalty*, is simply a percentage of the gross revenues received by a mining operation for the sale of its production, whether in the final market form or an intermediate product such as concentrate. These alternative names may contractually entail certain economic deductions, for example, marketing expense; however, in general, these are few, which make this royalty form very easy to audit.

In the present analysis, of the 1,932 royalties owned by the royalty companies, 187 (10%) were classified as GRR. Seventy-five percent pre-existed prior to purchase by the royalty company, with 63% acquired in portfolios.

The underlying projects represent 22 different mineral commodities, with gold being the largest (35%), and no more than 8% for any other individual mineral product. These assets are located in 21 countries with three jurisdictions representing 81% of the assets: Australia hosts the most (62), followed by Canada (59), and the United States (30). No other country has more than four GRRs.

There is a fairly uniform split on the distribution of GRRs by project status. Thirty-two percent involve projects at the prospecting stage, 27% reflect projects with resources and are in the process of completing engineering studies. Five percent are in development, and another 36% are attached to currently producing mines.

The gross royalty rates for 164 GRRs ranged from 0.07% to 27.5%, with a median value of 1.5% and a mean of 2.24%. Seventy-three percent of the gross royalties had rates less than 2.5%. In addition, there are 10 sliding-scale GRRs and another 13 with unknown terms (see the "Sliding-Scale and Variable Rate" section).

The three GRRs with rates above 20% represent price participation royalties wherein the royalty rate is tied to the spread between a set price and the realized spot metal price, multiplied by the units of production attributed to the royalty interest (Figure 1). This compares to the more common format of ignoring the spread and simply multiplying the production units by the spot price. The rates must be higher to compensate for the smaller spread range relative to the magnitude of the spot price. Despite being labeled as GRRs, these outliers more closely resemble offtake agreements.

Net Smelter Return

Producers that need to ship their mine product off-site to a third party for final processing, such as smelting and refining, have limited control over these external costs, so they prefer to deduct these from the revenues prior

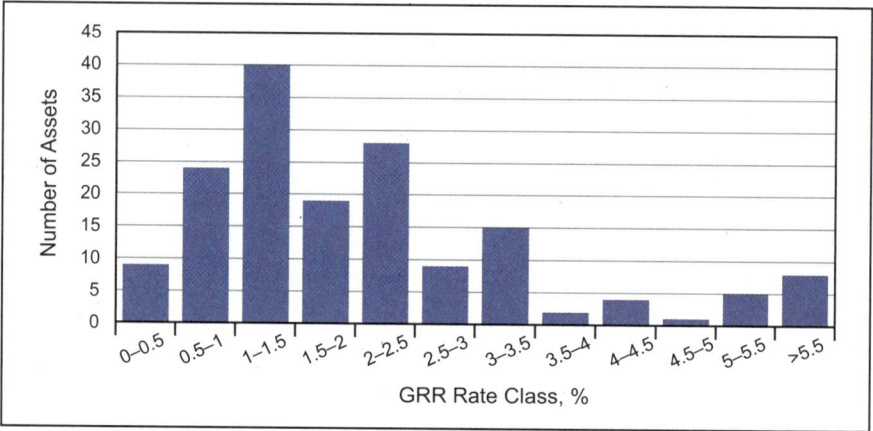

Figure 1 Distribution of gross revenue royalty (GRR) rates for 187 royalties

to making the royalty payment. This contractual arrangement is called a *net smelter return* or NSR. The deductions typically include expenses incurred for transportation, insurance, and smelting/refining costs. For gold, these deductions tend to total less than $10 per ounce, so in many ways, gold NSRs behave like GRRs. For base metal mines, these deductions can be quite substantial (upward to 30% of the gross value of the contained metal).

NSRs are the most common type of mining royalty. Of the 1,556 NSR royalties represented, 80% of the total royalties are owned by the mining royalty companies, with data available on 1,542. These represent 52 countries with Canada accounting for 64%. The United States is in distant second place with 373 (24%) NSR royalties. The next five locations by number are Mexico (70), Chile (61), Peru (56), Brazil (38), and Turkey (23). All other countries host less than 20 royalties, with most having fewer than 5.

Based on available data for 1,422 NSRs, their stipulated royalty rates range from 0.03% to 28%, with most less than 5% (Figure 2). The median value is 1.5%, and the modal range is between 1% and 1.5%. The uppermost value reflects a single large NSR on a secondary metal.

Based on available data, 15% of the NSRs were created by the royalty companies and 21% were purchased/created as single assets. The rest were pre-existing and acquired via portfolio purchases. More than 78% are focused

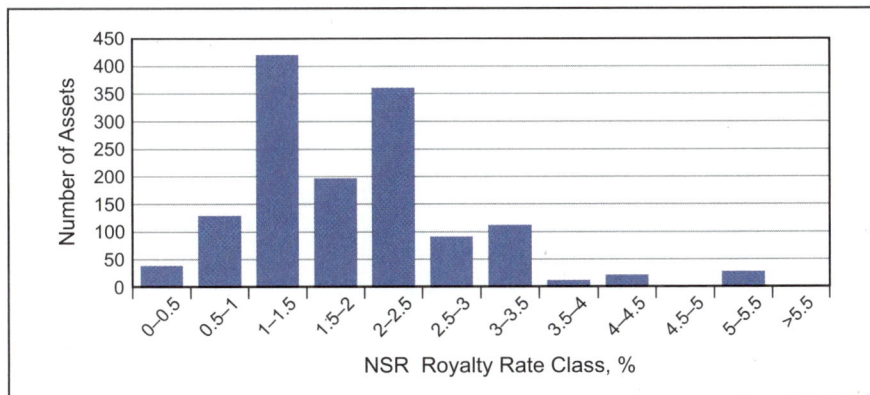

Figure 2 Distribution of net smelter return (NSR) royalty rates for 1,543 royalties

on gold, 9% on copper projects, 3% on nickel, and 2% on zinc. The other 19 commodities observed in NSR contracts have few representatives.

Fifty-six percent of the NSRs are on early stage prospects, 23% are linked to projects in the process of defining resources and conducting early engineering studies, 16% represent projects in development or production, leaving 6% attached to suspended, exhausted, or of unknown status operations. The large percentage of exploration-stage NSRs reflect the common practice of selling these high-risk, hence low-value, royalties in portfolios.

Net Profits Interest

A third form of royalty is the net profits interest (NPI) royalty (also referred to as a *net proceeds royalty* [NPR]). Operators prefer to pay a royalty only on net revenue after all of the operating, capital, and local tax costs have been deducted on the concept that the operator and royalty holder should be fully aligned and share in the mine's profits. The argument is that the royalty holder should not gain the benefit of excluding certain operating and capital costs that impact the mine's economics.

Royalty holders detest this type of royalty because it exposes them to cost inflation as well as accounting games that can be utilized to ensure a very small (or no) royalty is paid to the holder after permitted cost deductions. This is why an NPI is euphemistically known as "no proceeds intended." NPIs also require significant management time and resources to audit.

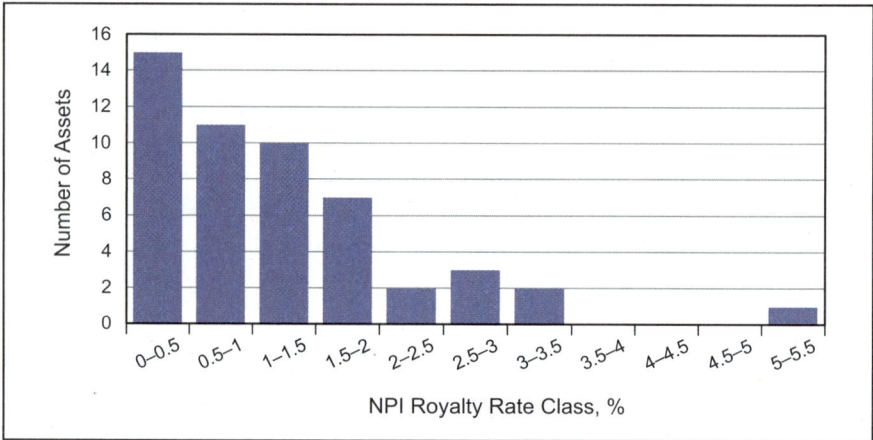

Figure 3   Distribution of net profit interest (NPI) royalty rates for 54 royalties

This data set contains 54 NPI royalties with rates ranging from 0.25% to 50% (Figure 3). The 50% NPI was payable after the operator had recovered all of its initial and sustaining capital costs. Fifty-one percent of the NPIs have rates between 5% and 15% with the 5%, 10%, and 15% interval classes representing most of the observations. The median NPI value was 10%, and there was only one slider. NPI royalty rates are typically multiples of the other royalty forms because of the substantial deductions involved in determining net profit.

NPIs are located in nine countries with Canada (21) hosting the most, followed by the United States (20), Australia (6), and Liberia (2). Ghana, Mexico, South Africa, Turkey, and Uganda each have one NPI royalty.

Seventy-four percent of this set are for gold. Copper (5) and uranium (4) represent the next 17%, followed by one occurrence each for silver, cobalt, lithium, nickel, and platinoids. Thirty-five percent of the NPIs are on prospecting-level assets with another 25% of projects currently at the engineering stage. Thirty percent represent development and producing projects.

More than 90% of the NPIs were pre-existing and purchased in portfolios by the royalty companies. Again, this reflects their relative undesirability.

Table 1 Relationship between the income statement and royalty type

Income Statement	Royalty Type
Gross revenues	Gross revenue royalty
Less smelting, refining, transportation, insurance expenses	
Net smelter return	Net smelter return royalty
Less operating expenses	
Less capital costs	
Less local taxes	
Net income	Net proceed interest royalty

The differences in how revenue due to a royalty owner is calculated under the three royalty forms are illustrated in Table 1 by way of an income statement. As one would expect, GRRs involve the fewest deductions from gross revenue while NPI royalties have the most.

Fixed

A fourth type of royalty is commonly used for bulk minerals such as industrial minerals, sand and gravel, iron ore, or coal. These mineral products can be difficult to establish market-based prices for, so a set charge per metric or short ton mined, processed, or sold is employed. This construct is called a *fixed royalty*. An example is construction sand where the mineral owner receives $0.75 per ton of material extracted from its property. Today, fixed royalties are also observed for base and precious metals projects with unit royalties set on a per metric ton, ounce, or pound basis. Fixed royalties are less desirable to royalty companies because of their low optionality; the owner cannot participate in market upsides. In this study, there are 81 fixed-rate royalties involving 14 different commodities. Fifty royalties were for gold, followed in descending order by coal (6), iron (5), and copper (4). The balance includes potassium (2), nickel (2), uranium (2), zinc (2), and one each for aluminum, lead sand, sulfur sand, and zeolites. Two were for unknown commodities.

For this study, discrete cash payments made by the mine's operator to the royalty holder were classified as fixed royalties. An example would be a fixed payment of $2 million due when the mine was commissioned, with no other terms requiring future financial outlays.

Annual minimum royalty. Another form of a fixed royalty is an annual minimum royalty (AMR). AMRs are effectively rental payments made to the royalty holder while the project is not in production. Once production commences, the amounts paid to date for the AMR are commonly recaptured by the mine operator/owner out of future royalty payment obligations.

These fixed royalties occur for projects in 12 countries. Australia has the most (42), followed by Canada (18), the United States (9), and Turkey (3). There are two in Argentina and one each in Armenia, Brazil, Cameroon, Kazakhstan, Mexico, Namibia, and the Philippines.

Only four fixed-rate royalties were created by royalty companies. At least 39 were pre-existing with 11 having unknown status. Six were purchased as single assets, 50 as components in portfolios, and 25 are of unknown origin.

A summary of the distribution of the various royalty structures is presented in Table 2.

As the following chapters show, the mining royalty industry has become a major financing source to the mining industry. This has led to the invention of hundreds of variations on these four basic royalty themes. Four of the more common variations are described next.

Sliding-Scale and Variable-Rate

Sliding-scale and variable-rate royalties (*sliders*) pay one rate under certain defined conditions and a different (usually higher) rate under changed conditions. Most sliders are attached to commodity prices. For example, a royalty rate is set at 1% for gold sales prices of $1,000 per ounce and under, 1.5%

Table 2 Distribution of royalty structures

Structure	Abbreviation	Number	Percentage, %
Net smelter return royalty	NSR	1,556	80
Gross revenue royalty	GRR	187	10
Fixed royalty	Fixed	81	4
Net profits interest royalty	NPI	54	3
Unknown	—	54	3
Total		**1,932**	**100**

when the gold price is between \$1,000 and \$2,000 per ounce, and 2% when the gold price exceeds \$2,000 per ounce. Sliders can also be linked to mine capital repayment schedules, production milestones, or throughput expansion rate achievements. In one case, different royalty rates were tied to the composition of ores being processed. Investors like slider royalties because the royalty holder benefits from higher commodity prices or increased throughput rates via a concurrent higher royalty rate.

In the examined data set, there are 142 sliders. Sliding-scale NSRs comprise the greatest number with 125, followed in descending order by 10 GRRs, 4 fixed, and 3 NPIs. The NSR sliders rates range from downside conditions of 0% to maximums of 28%, with 1% as the average spread between the high and low royalty rates (Figure 4). Thirty-eight percent of the NSR sliders work within this spread range, and 72% have spreads of less than 1.5%.

Only seven NPI sliders are in this data set. The rates range from 2.3% to 15%, and the spread is between 3.6% and 7.5%. The larger spread reflects the need for greater economics, given the large number of deductions taken by the operator.

Step-Up or Step-Down

Step-up or step-down royalties are becoming more popular. The royalty holder accepts a lower royalty rate until the mine's capital is paid off and then

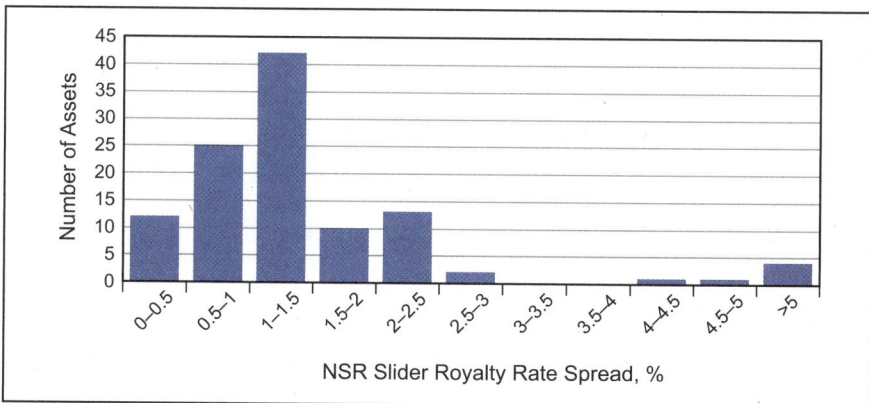

Figure 4 Distribution of sliding-scale net smelter return (NSR) royalties' spread rates for 125 royalties

receives a higher rate for the royalty's remaining term. A final approach is to cap the royalty rate when a certain amount of royalty payments has been made and then step down the rate for the remaining mine's life. An example would include a 2% NSR until $10 million has been paid out and then the rate reduces to a 1% NSR for the mine's duration. The transition point is often related to that point where royalty holders have captured back their initial investment plus some rate of return and is driven by the operator's desire to limit the upside.

Royalty companies prefer not to have their royalties capped as the royalty transaction then looks more like a loan. Generally, royalty companies do not insist on high royalty rates because it is to their advantage not to be a meaningful factor to the mine's economics. If it is, it runs the risk of negatively impacting the mine plan and/or gives the operator the opportunity to ask for a reduction during times of low metal prices or operational challenges. By keeping the royalty rate low, the royalty holder can often make a strong case that there are other, more impacting ways to reduce the mine's costs and keep it operating.

Streams

Commodity streaming arrangements involve the payment of an up-front amount to the mine developer/operator for the right to purchase one or more of the mine's products over a set period under predefined terms. The streaming payment is made when the product is delivered to market, and its price is typically the lesser of the spot market price and a price set out in the streaming contract. Most streams also have an inflation adjustment factor (typically 1% per year) that kicks in a year or two after production commences.

Streams were originally designed to arbitrage the value of the precious metals co- and by-products recovered at base metal mines. Because base metal operations typically have longer lives than primary precious metals–producing mines, this also makes streams very attractive to the royalty companies.

The principal difference between streams and royalties is size. While a large NSR royalty might be about 2% attached to all metal production from the mine, streams can be for up to 100% of the by-product gold and/or silver production, and hence could represent a substantial portion of the mine's revenues.

The up-front and ongoing payments are linked to the net present value of the instrument, reflecting what the stream purchaser pays up front and the contract-stipulated price it will pay per metal unit for future deliveries. If a development project is in the construction phase and requires substantial funds to cover evolving capital costs, the trade-off is giving the stream purchaser lower ongoing payment terms for the contracted commodity in exchange for higher initial capital investment. This is why it is common to see low ongoing payment rates relative to the spot metal prices observed at time of agreement, which provides the royalty investor with more leverage in a rising price scenario.

As illustration, Silver Wheaton Corp. (the predecessor company to Wheaton Precious Metals Corp.) innovated the streaming concept in 2004 and in early agreements set the initial unit purchase prices at the lesser of $400 per ounce of gold/$3.90 per ounce of silver and the contemporary spot prices. For reference, spot market values at the time were approximately $440 per ounce for gold and $7.00 per ounce for silver.

The set prices worked well when precious metal prices were low. However, today's prices are much higher so it becomes penurious to the operator to pay a low set price. Royal Gold realized this in 2010 when negotiating a stream on the Andacollo (Chile) mine and switched to a percentage of the spot price for determining the ongoing purchase payment it would make for the contracted production. With precious metals prices significantly increasing since 2004, there has been a logically increasing preference for percentage-based ongoing payments over fixed rates, as shown in Figure 5.

As shown in Figure 6, since 2004, approximately 7 streaming transactions have been closed each year, with 2017 being a banner year with 16. On average, about $1.3 billion has been spent each year in streaming deals by the royalty companies. An exceptional year was 2015 with $6.1 billion expended.

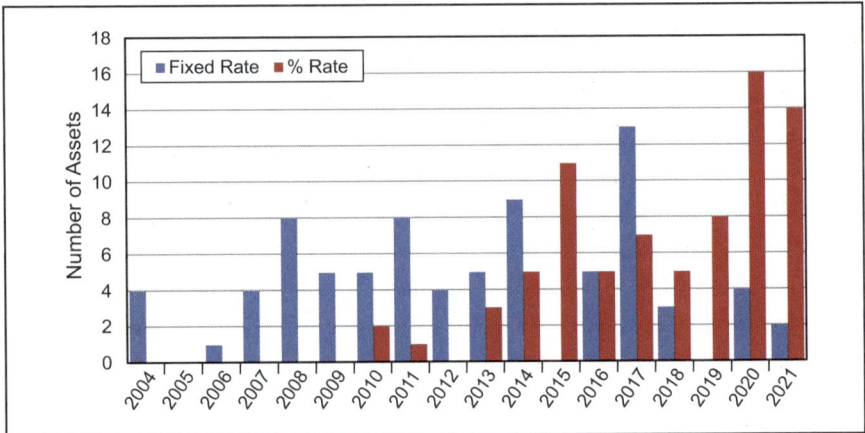

Figure 5 Distribution of fixed-rate versus percentage-rate ongoing payments

Wheaton Precious Metals may have invented the streaming model, but the instrument's popularity has clearly increased each year. Based on available data, more than $23 billion has been spent on streams since 2004, including 4 transactions in excess of $1 billion and 16 valued over $500 million.

One hundred and thirty-five streaming agreements are held by 15 royalty companies with Wheaton Precious Metals controlling almost 30%. The streams are linked to 10 metal types with precious metals representing by far the most numerous at 92%. The balance involves copper, cobalt, diamonds, nickel, platinum, uranium, and zinc. Fifty-four percent of the streams are linked to the mine's co- and by-product metals. More than 90% of the streams are on projects either in development or production.

Most commodity offtakes and conventional royalties acquired over the past two decades were pre-existing and purchased as components in port-folios. However, for the new model of metal stream financings, 70% of the streams were created by the royalty companies, and 73% represented single-asset investments.

Geographically, streaming agreements are attached to projects in 28 coun-tries, with Canada (35), Mexico (16), United States (14), Peru (10), and Brazil (8) representing the top five countries and receiving 61% of the royalty company investments. Other countries with stream-attached activities

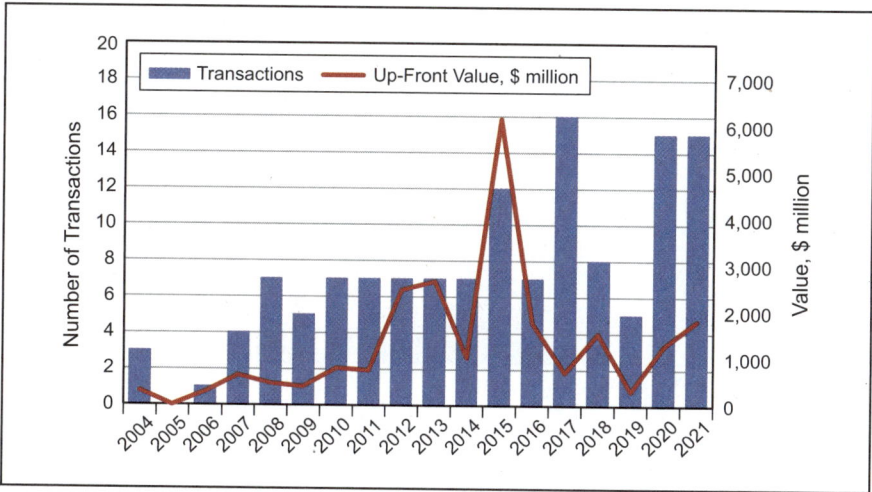

Figure 6 Stream transactions chronology

are Chile (7), Australia (6), South Africa (5), Mongolia (4), Panama (4), Argentina (3), Burkina Faso (2), Colombia (2), Ghana (2), Macedonia (2), Portugal (2), and one each in Armenia, Botswana, Côte d'Ivoire, Dominican Republic, Fiji, Greece, Guyana, Honduras, Nicaragua, Senegal, Sweden, and Tanzania.

As previously noted, Wheaton Precious Metals originated and derives most of its revenues from streams. Franco-Nevada Corporation realized its first stream revenues in 2010 and, as shown in Figure 7, streams now account for approximately 70% of the sector's total revenues.

The mechanisms and metrics of metal streaming contracts can be challenging to understand for the uninitiated. This is particularly true in trying to assess the differences in value between the various royalty types to both the royalty owner/creator and the mine owner/operator. To assist, Franco-Nevada Corporation shares an example that "compares a 4% NSR to a 4% stream, NPI or WI (working interest)."[1] Assume that for 1 ounce of gold, the realized sale price is $1,800 per ounce at time of delivery. The "stream cost" is $400

1. Franco-Nevada Corporation, *Asset Handbook 2022* (Toronto: Franco-Nevada, 2022), 34, used with permission.

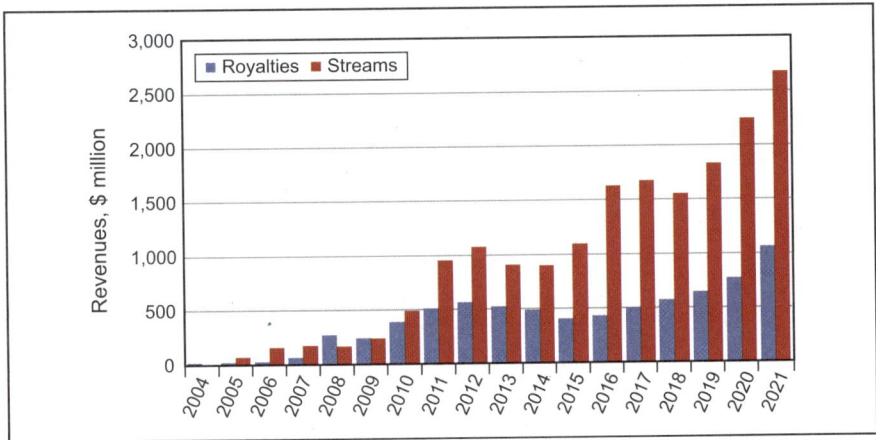

Figure 7 Revenue breakdown for the 15 public royalty companies with streams

per ounce and the "all-in sustaining cost" of the mine is $1,026 per ounce, as depicted in Table 3.

Franco-Nevada continues: "Based on the above economics, a comparable percentage NSR is 2.3 times more valuable than an equivalent developed NPI or WI and almost 1.3 times more valuable than a stream interest. The NSR provides the highest margins and most downside protection to changes in the commodity price. The stream provides commodity price leverage similar to a low-cost operating company with certainty as to future costs. The NPI or WI provides the most leverage to commodity prices."[2]

Offtakes

Offtakes are contractual agreements in which the buyer agrees to purchase a certain amount of the mine's product output over time under negotiated terms. These contracts are the mainstay business of metal traders who profit from the spread between the set purchase price of the commodity and sales price when marketed. Traders also employ numerous hedging strategies to boost their profit margins using the offtake contracts as collateral.

---

2. Franco-Nevada Corporation, *Asset Handbook 2022*, Toronto: Franco-Nevada, 2022), 34, used with permission.

Table 3 Comparison of net smelter return royalties to streams, net profits interest, and working interest

Item	Net Smelter Return (NSR)	Stream	Developed Net Profits Interest (NPI) or Working Interest (WI)*
1 ounce sold	$1,800	$1,800	$1,800
Applicable cost	$0	$400	$1,026
Margin for all-in sustaining cost (AISC) calculation	$1,800	$1,400	$774
NSR, stream, or NPI	4%	4%	4%
Revenue per ounce to Franco-Nevada Corporation (FNV)	$72	$56	$31
NSR equivalent	100%	78%	43%
Alternatively, ounces required to equal a 1% NSR	1 oz	1.3 oz	2.3 oz

* For applicable costs for a developed NPI or WI, FNV is, for illustrative purposes, assuming Barrick Gold Corporation's 2021 AISC measure as Barrick is the operator of two assets at which FNV has NPI interests.

One large difference between offtakes and streams is that offtake agreements price their product based on the spread between two prices over a period. For example, the offtaker selects the daily price for the prior 30 days. In a streaming arrangement, the streamer is required to pay the lesser of the spot price and a fixed amount.

In the study, 18 offtake contracts were identified, which are owned by three royalty companies. Contracted output rates range from 25% to 100% of production. Seventeen represent gold contracts and one is for silver. One offtake is for by-product gold production from a copper mine, but the rest involve the mine's primary commodity.

All of the observed offtake agreements represented pre-existing contracts, in place before acquisition by the royalty companies. Only one contract was purchased as a single mineral asset. The rest were purchased as part of portfolios consisting of offtakes, royalties, and/or streams.

The projects subject to these offtake commitments are located in nine countries. Mexico has the most with four, followed by three each in Brazil,

Canada, and the United States. The balance has one contract each in Armenia, Australia, Côte d'Ivoire, South Africa, and Turkey.

Offtakes are most commonly applied to currently producing or near-production assets. At the time of contract formation, nine of these offtakes were on producing operations, three on development-stage projects, one at feasibility-stage status, three at preliminary economic assessment stage, and two still at the delineation stage.

All of these contracts were entered into with small- to medium-size mining and exploration companies because the larger organizations typically have their own marketing departments that deal directly with the metal traders. Their strong financial positions and downstream market involvement preclude the need to sell an offtake to a royalty company.

INDUSTRY STATISTICS

At the end of December 2021, the 36 examined royalty companies owned 2,085 assets consisting of 18 offtakes, 1,932 royalties, and 135 streams. The project status of each of these assets is presented in Table 4. Royalty companies are prone to obtaining early stage assets, motivated by either the low-entry cost or because they are packaged in a portfolio with other higher-quality assets. Notably, more than half these assets involve projects without any currently delineated resources. By contrast, 75% of the streamers' assets are on near-term production opportunities.

Collectively, these categorical structures reflect 31 different commodities. Based on available data, 14% of these royalties were created by the existing royalty companies, and 86% were pre-existing when transacted. Seventy-eight percent were acquired as components of portfolio purchases.

These early stage royalties are typically created along the following lines: An exploration company (optionee) options an early stage prospect from the property owner (optionor). To earn its equity interest, it has to either spend a certain amount of money or advance the project to a certain delineation or development stage over a set period. For option agreements where the

Table 4 Distribution of offtakes, royalties, and streams by project status

Project Status	Offtakes	Royalties	Streams	Total	Percentage, %
Prospecting	0	931	5	936	44.9
Discovery	0	40	0	40	1.9
Delineation	2	242	2	246	11.8
Preliminary economic assessment/prefeasibility	3	120	3	126	6
Feasibility	0	73	5	78	3.7
Bankable feasibility	0	11	4	15	0.7
Development/construction	3	86	12	101	4.8
Producer	10	312	86	408	19.6
Suspended	0	41	18	59	2.8
Unknown	0	76	0	76	3.6
Total	**18**	**1,932**	**135**	**2,085**	**100**

optionee can earn 100% interest or where the owner's interest is diluted below 10%, the interest is typically converted into a royalty. The rationale for this conversion is that the minority owner has no control over project spending but wants to retain an economic interest in the potentially economic asset. By converting to a royalty, the minority owner doesn't have to invest further capital for exploration or development expenses but receives economic benefit should production occur in the future.

The problem in tracking these exploration royalty conversions is that exploration companies don't know the timing to achieve first production nor the project's overall economic value at the time of conversion. This causes the newly formed royalty to have very little contemporary value. Therefore, the conversion from an equity to a royalty interest is usually not material to either party. Documentation is often rudimentary or obtuse with no disclosure of creation date without full title research, making it difficult, if not sometimes impossible, to track the royalty's creation date. The challenge also reflects the large number of royalties created by small prospectors; most which will not be in production for years, if ever.

Most royalty companies strive to purchase assets involving advanced-stage projects. However, 53% of the assets in this study were acquired in portfolios where the seller combined different project-stage assets into one package. This is reflected by the fact that 59% of the assets were pre-existing royalties. While 100% of the offtakes were pre-existing, 70% of the streams were created.

On a regional level, North America hosts 61% of the assets analyzed. Oceania, overwhelmingly represented by Australia, represents 15%, followed by South America with 12%. The preference for these geographies is driven by stable governments and strong mining laws.

The top eight countries for offtakes, royalties, and streams constitute 87% of the assets in the data set. Canada represents 35% of the total, followed in descending order by the United States (22%), Australia (15%), Mexico (5%), Brazil (3%), Chile (3%), Peru (3%), and Argentina (2%). The next nine largest countries each represent 1% of the data set, bringing the total to 94%, and include Burkina Faso, Côte d'Ivoire, Ghana, Mongolia, Norway, Serbia, South Africa, Sweden, and Turkey.

Canada hosts the world's most populated mining stock exchanges; it is home to spectacular economic geology and mineral potential, and local familiarity with the rocks, culture, politics, and regulations make projects in this jurisdiction more desirable. These virtues are also a key reason for the Canadian-centric nature of the global junior mining sector.

Timelines

A review of the purchase dates for these assets by the royalty companies active across the study period is shown in Figure 8. Since 2007, there has been a substantial increase in acquisition activities by a growing number of firms entering the non-petroleum royalty industry, with a new record set in 2021.

Sixty percent of these assets represented pre-existing contract agreements with royalties representing 95% of the total (Table 5). By contrast, only 14% of the observed arrangements were originated by the royalty companies them-selves, as opposed to existing contracts. Again, royalties comprised the largest category, representing 66% of the new contractual creations.

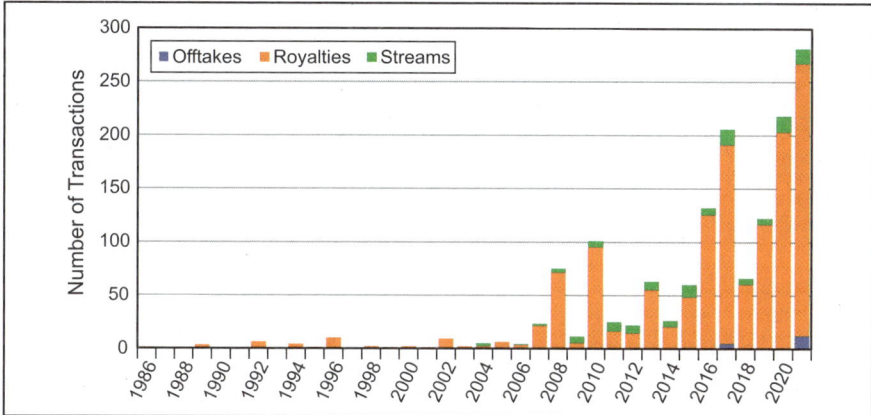

Figure 8 Number of offtake, royalty, and stream transactions by active royalty companies

Table 5 Distribution of created versus purchased royalty company assets

Type	Created	Pre-Existing	Unknown	Total
Royalties	187	1,192	553	1,932
Streams	94	40	1	135
Offtakes	1	18	0	19
Total	282	1,250	554	2,086
Percentage, %	14	60	26	100

One might conclude that there was an onslaught of pre-existing assets put on the market in the later years of the study period, but the numbers belie the fact that more than half of the individual purchases were contained in portfolios, as summarized in Table 6.

The increase in royalty acquisition activity began during the Chinese supercycle during the first decade of the 21st century, when investors moved back into the mining industry as a response to rising commodity prices. This opened the door for more royalty companies to be formed. Figure 9 shows the increasing levels of annual acquisition activity. The higher spikes reflect the acquisition of large portfolios, usually via corporate mergers and takeovers of competitors.

Table 6 Single-asset versus portfolio acquisitions by existing royalty companies

Type	Single Asset	Portfolio	Unknown	Total
Royalties	302	1,080	550	1,932
Streams	96	39	0	135
Offtakes	2	17	0	19
Total	**400**	**1,136**	**550**	**2,086**
Percentage, %	**19**	**54**	**26**	**100**

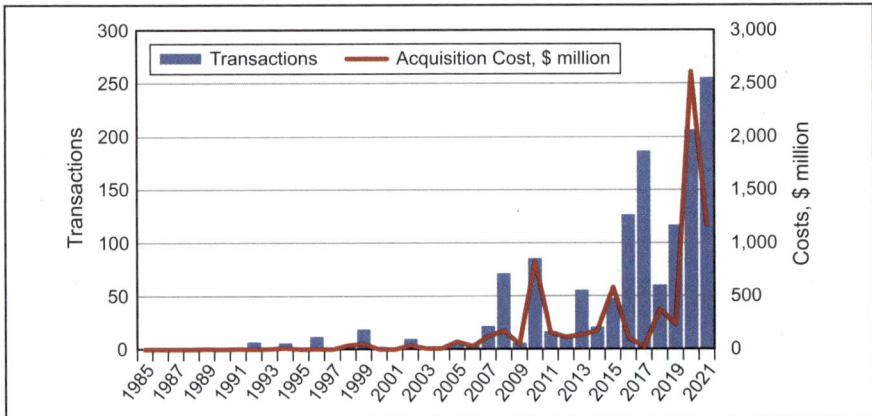

Figure 9 Acquisition timeline for royalty transactions by royalty companies

Based on the data available for the study, an estimated $7.2 billion has been spent on royalty acquisitions since 1986 by the sector companies. This level of outlays is almost a third of what has been spent on streams. (Note: Many transactions, especially involving exploration-stage projects, are not included in these estimates because of lack of data.)

Eighty-eight percent of the assets and 91% of the cumulative transaction value occurred since 2010. Much of the recent activity is related to royalty companies acquiring each other. For example, over the study period, 18 public royalty companies have been acquired by their competitors for a total of $7.4 billion (Table 7). The four corporate transactions involving Franco-Nevada as either a buyer or seller represents $4.8 billion (65%) of this total. This activity includes two non-royalty company buyers, Newmont Corporation and IAMGOLD Corporation, which expended $3.1 billion (46%) for royalty assets. In addition, there are a substantial number of private

Table 7 Public mining royalty company takeovers

Purchased Company	Buyer	Acquisition Year	Amount, $ million
Redstone Resources	Franco-Nevada Mining Corporation	1996	65
Euro-Nevada Mining Corporation Limited	Franco-Nevada Mining Corporation	1999	1,031
High Desert Mineral Resources	Royal Gold	2002	30
Franco-Nevada Mining Corporation	Newmont Corporation	2002	2,900
Repadre Capital Corporation	IAMGOLD Corporation	2003	214
Battle Mountain Gold	Royal Gold	2007	47
Silverstone Resources Corporation	Wheaton Precious Metals Corp.	2009	197
International Royalty Corporation	Royal Gold	2010	728
Gold Wheaton Gold Corporation	Franco-Nevada Corporation	2011	831
Vaaldiam Mining	International Royalty Corporation	2012	19
Premier Royalty	Sandstorm Gold	2013	27
Sandstorm Metals & Energy	Sandstorm Gold	2014	59
Gold Royalty Corporation	Sandstorm Gold	2015	5
Callinan Royalties Corporation	Altius Minerals Corporation	2015	92
Terraco Gold Corporation	Sailfish Royalty Corporation	2019	17
Ely Gold Royalties	Gold Royalty Corporation	2021	192
Abitibi Royalties	Gold Royalty Corporation	2021	231
Golden Valley Mines and Royalties	Gold Royalty Corporation	2021	118
Total			**6,803**

royalty companies for which information is nearly impossible to track because of lack of public disclosures or non-materiality of assets.

The Companies

The study identified 36 active public mining royalty companies as of the end of 2021, and another 18 that were in operation but taken over during 1986 through 2021, for a total of 54 individual royalty firms that existed at some point or continue operating today (Table 8).

Table 8 List of mining royalty companies over time

Company Name	Status	Stock Exchanges*	Stock Symbol	Royalty Company Formed	Company Exit Year
Abitibi Royalties	Inactive	TSXV, OTCQX	RZZ, ATBYF	2015	2021
Altius Minerals Corporation	Active	TSX, OTCQX	ALS, ATUSF	2003	—
Altus Strategies	Active	AIM, TSXV, OTCQX	ALS, ALTS, ALTUG	2019	—
Anglo Pacific Group	Active	LSE, TSX	APF, APY	2001	—
Battle Mountain Gold Exploration	Inactive	OTCBB	BMGX	2004	2007
Callinan Royalties Corporation	Inactive	TSX	CAA	2011	2015
Deterra Royalties Limited	Active	ASX	DRR	2020	—
Electric Royalties	Active	TSXV, OTCQB	ELEC, ELECF	2020	—
Elemental Royalties Corporation	Active	TSXV, OTCQX	ELE, ELEMF	2020	—
Ely Gold Royalties	Inactive	TSXV, OTCQX	ELY, ELYGF	2016	2021
Empress Royalty Corporation	Active	TSXV, OTCQB	EMPR, EMPYF	2020	—
EMX Royalty Corporation	Active	TSXV, NYSE AM, FRA	EMX, EMX, 6E9.F	2017	—
EURO Ressources	Active	NYSE Euronext (PAR)	EUR	2004	—
Euro-Nevada Corporation	Inactive	TSX	EN	1987	1999
Franco-Nevada Corporation	Active	TSX, NYSE	FNV	2007	—
Franco-Nevada Mining Corporation	Inactive	TSX	FN	1986	2002
Gold Royalties Corporation	Inactive	TSXV	GRO	2012	2015
Gold Royalty Corporation	Active	NYSE AM	GROY	2020	—
Gold Wheaton Gold Corporation	Inactive	TSX	GLW	2008	2011
Golden Valley Mines and Royalties	Inactive	TSXV, OTCQX	GZZ, GLVMF	2021	2021
Great Bear Royalties Corporation	Active	TSXV	GBRR	2020	2022

continues

Table 8 List of mining royalty companies over time (continued)

Company Name	Status	Stock Exchanges*	Stock Symbol	Royalty Company Formed	Company Exit Year
High Desert Mineral Resources	Inactive	TSXV	HDR	2001	2002
International Royalty Corporation	Inactive	TSX, AMEX	IRC, IRC	2004	2010
Labrador Iron Ore Royalty Corporation	Active	TSX, OTC PINK	LIF, LIFZF	1995	—
Maverix Metals	Active	TSX, NYSE AM	MMX, MMX	2016	—
Mesabi Trust	Active	NYSE	MSB	1961	—
Metalla Royalty & Streaming	Active	NYSE, TSXV	MTA, MTA	2016	—
Morien Resources Corporation	Active	TSXV, OTC PINK	MOX, APMCF	2014	—
Mundoro Capital	Active	TSXV, OTCQB, FRA	MUN, MUNMF, NGU	2019	—
Natural Resource Partners	Active	NYSE	NRP	2002	—
Nickel 28 Capital Corporation	Active	TSXV	NKL	2019	—
Nomad Royalty Company	Active	TSX, OTCQX, FRA	NSR, NSRXF, IRLB	2020	2022
Nova Royalty Corporation	Active	TSXV, OTCQB	NOVR, NOVRF	2020	—
Orogen Royalties	Active	TSXV, OTCQX	OGN, OGNRF	2020	—
Osisko Gold Royalties	Active	TSX, NYSE	OR, OR	2014	—
Premier Royalty	Inactive	TSX	NSR	2012	2013
Redstone Resources	Inactive	TSX	RR	1989	1996
Repadre Capital Corporation	Inactive	TSX	RPD	1992	2003
Royal Gold	Active	NASDAQ	RGLD	1991	—
Sailfish Royalty Corporation	Active	TSXV, OTCQX	FISH, SROYF	2017	—
Sandstorm Gold	Active	TSX, NYSE	SSL, SAND	2009	—

continues

Table 8 List of mining royalty companies over time (continued)

Company Name	Status	Stock Exchanges*	Stock Symbol	Royalty Company Formed	Company Exit Year
Sandstorm Metals & Energy	Inactive	TSXV	SND	2010	2014
Scully Royalty	Active	NYSE	SRL	2019	—
Silverstone Resources Corporation	Inactive	TSXV	SST	2007	2009
Star Royalties	Active	TSXV, OTCQX	STRR, STRFF	2018	—
Terraco Gold Corporation	Inactive	TSXV	TEN	2011	2019
Trident Royalties	Active	AIM	TRR	2020	—
Triple Flag Precious Metals Corp.	Active	TSX, TSX	TFPM, TFPM.U	2019	—
Uranium Royalty Corporation	Active	TSXV, NASDAQ	URC, UROY	2017	—
Vaaldiam Mining	Inactive	TSX	VAA	2010	2012
Vox Royalty Corporation	Active	TSXV, OTCQX	VOX, VOXCF	2020	—
Voyageur Mineral Explorers Corporation	Active	CSE	VOY	2021	—
Wheaton Precious Metals Corp.	Active	TSX, NYSE, LSE	WPM, WPM, WPM	2004	—
Xtierra	Active	TSXV	XAG	2019	—

* AIM = Alternative Investment Market (London Stock Exchange junior exchange); AMEX = American Stock Exchange; ASX = Australian Securities Exchange; CSE = Canadian Securities Exchange; FRA = Frankfurt Stock Exchange; LSE = London Stock Exchange; NASDAQ = National Association of Securities Dealers Automated Quotations; NYSE = New York Stock Exchange; NYSE AM = New York Stock Exchange American; NYSE Euronext (PAR) = New York Stock Exchange Euronext Paris; OTCBB = Over-the-Counter Exchange Bulletin Board; OTC PINK = Over-the-Counter Exchange Open Market; OTCQB = Over-the-Counter Exchange Venture Market; OTCQX = Over-the-Counter QX Exchange; TSX = Toronto Stock Exchange; TSXV = Toronto Stock Exchange Venture Exchange.

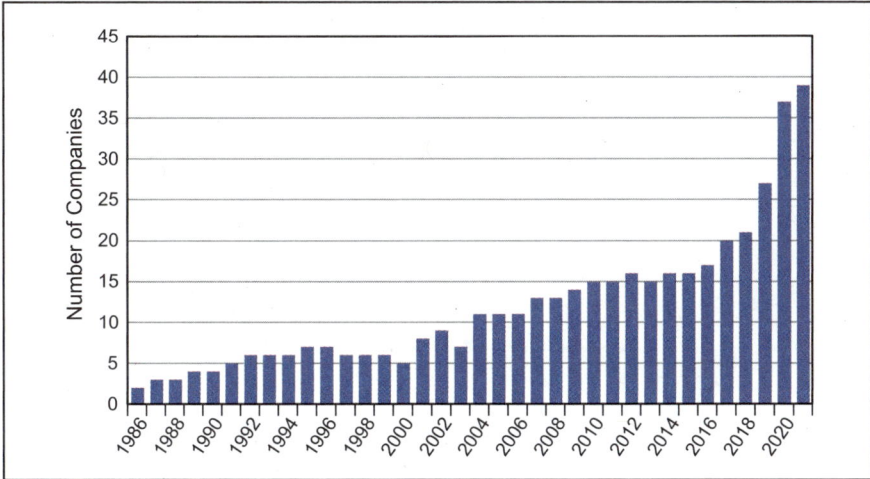

Figure 10 Growth of the public mining royalty sector

Mesabi Trust, formed in 1961, is the oldest mining royalty-type company. Mesabi's mission is to pay dividends, and they have never made efforts to expand their original portfolio holdings. Franco-Nevada Mining Corporation began its quest to accumulate mineral royalties in 1986. In that year, these two public royalty companies had a combined market capitalization of $42 million.

During the next 17 years (1986–2003), there were fewer than 10 public mining royalty companies at any point. The surge in new companies commenced around the start of the Chinese supercycle in 2004 (Figure 10). Three companies were taken over during 2021, and currently, several more appear to have become prey to the larger companies.

Seventy-nine percent of these companies have Canadian domicile. Of the companies, four (8%) are in the United States, three (6%) are in the United Kingdom, and one each (2%) is headquartered in Australia and France. The last two companies are domiciled in the British Virgin Islands and the Cayman Islands. Of the 18 defunct companies, 17 were Canada based and the other was in the United States.

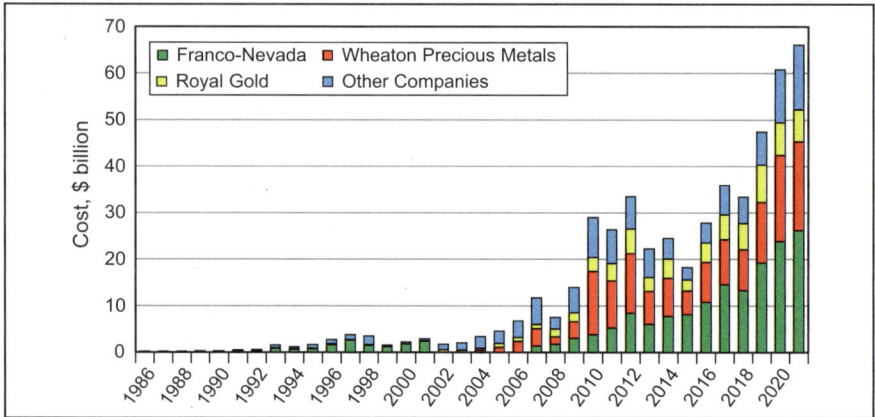

Figure 11   Total year-end market capitalization of analyzed royalty companies

Today's active companies have primary listings across nine stock exchanges, with the Toronto Stock Exchange Venture Exchange (TSXV) hosting 14, followed by the TSX (10), New York Stock Exchange (NYSE; 4), alternative investment market of the London Stock Exchange (LSE/AIM; 3), and one each on the Australian Securities Exchange (ASX), Canadian Securities Exchange (CSE), NASDAQ, and NYSE Euronext.

The industry's total market capitalization exceeded $66 billion at the end of 2021, with Franco-Nevada Corporation, Wheaton Precious Metals, and Royal Gold representing 79% of this value (Figure 11). The massive increase in market capitalizations began around 2003 when the number of public companies first exceeded 10.

There are four natural divisions between market capitalization sizes (Figure 12). The three Tier 1 companies (Franco-Nevada Corporation, Wheaton Precious Metals, and Royal Gold) each have market capitalizations in excess of $5 billion and as noted constitute 79% of total industry value.

The five Tier 2 companies (Osisko Gold Royalties, Labrador Iron Ore Royalty Corporation, Triple Flag Precious Metals Corp., Deterra Royalties Limited, and Sandstorm Gold) have individual market capitalizations ranging from $1 to $3 billion, accounting collectively for $8.5 billion (13%) of the industry's total market capitalization.

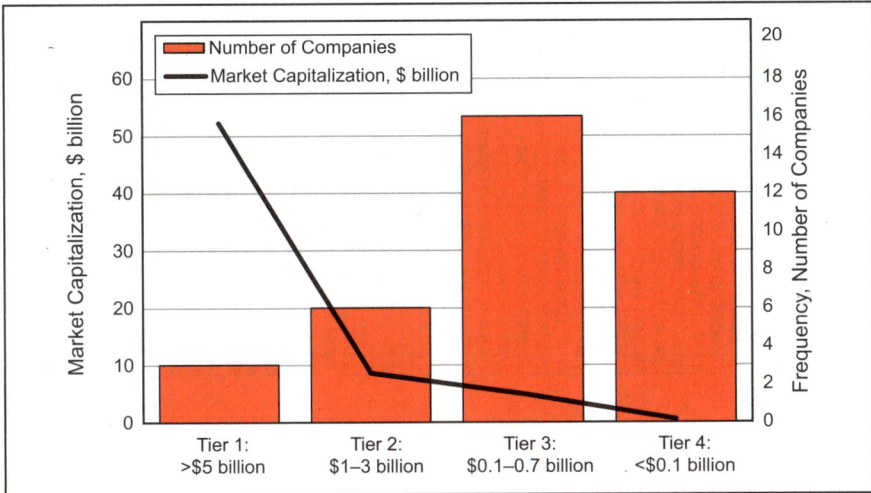

Figure 12 Year-end 2021 total market capitalization by tier class

Companies in the Tier 2 group should be of the most interest to investors because of the tightness in their relative market values. With a tight spread of just over $800 million between the largest (Osisko Gold Royalties at $2.1 billion) and smallest (Sandstorm Gold at $1.2 billion), the five companies making up this group tend to leapfrog each other as they compete to be the fourth largest royalty company and capture that coveted position among institutional investors.

Tier 3 royalty companies represent the most companies (16), and each firm has a market capitalization between $100 and $700 million. Collectively, these companies represent about $5 billion (8%) in sector market value. The study identified four companies in Tier 4, each with market capitalizations of under $100 million.

It is interesting that the three largest companies have been in existence for at least 17 years (Franco-Nevada Corporation being considered an extension of Franco-Nevada Mining Corporation. This provided these early players with significant competitive advantages in the market in name recognition, size, and most importantly, in well-established banking relationships.

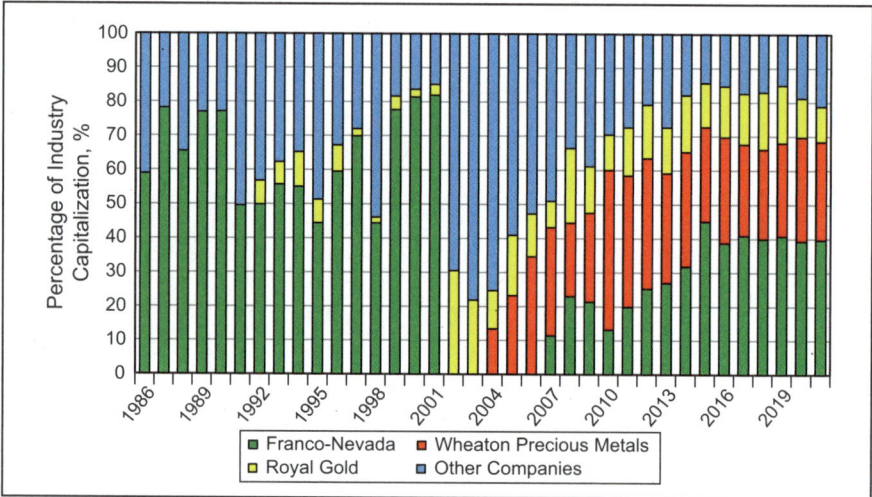

Figure 13 Comparative share of 2021 market capitalization for analyzed royalty companies

By contrast, 61% of the Tier 3 and 4 companies have been around for less than five years, trying to belatedly mimic the business models developed by the Tier 1 innovators.

Figure 13 shows the industry's composite 2021 market capitalization segregated by the three largest companies versus the collective smaller ones. The commanding position of Franco-Nevada Mining Corporation between 1986 and 2001 is obvious. This company was then merged into Newmont Corporation in 2002, effectively removing it from the public mining royalty industry. Around this time, Royal Gold was emerging as a significant mining royalty player. In 2004, Wheaton Precious Metals (under its predecessor name of Silver Wheaton) joined the industry. Franco-Nevada Corporation emerged from Newmont in 2007 as a spin-out and retook its leadership position. Today, there are a host of new participants, but the same three Tier 1 companies continue to dominate the industry.

This market capitalization growth is a direct function of the industry's revenue growth. In 1986, the industry's annual revenues were less than $2 million (Figure 14). By 2021, this revenue had increased to $4.5 billion. Of significant

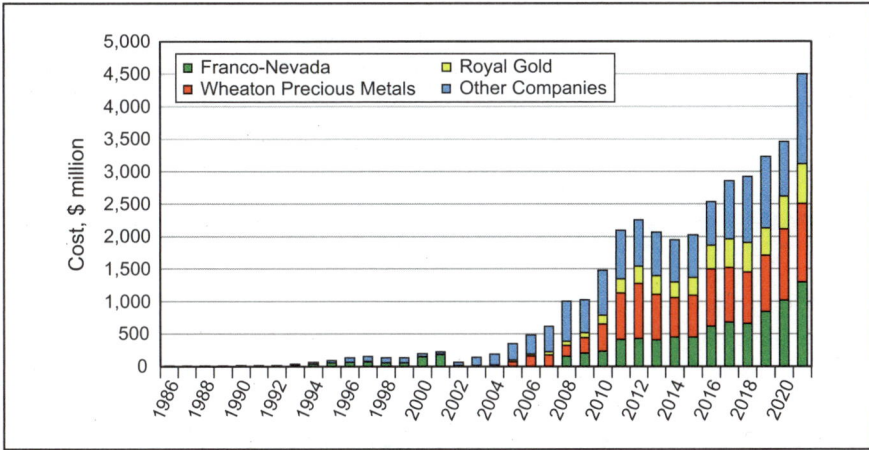

Figure 14 Annual industry revenues for analyzed royalty companies

note is the pronounced 30% growth in total sector revenues between 2020 and 2021.

Of the 408 revenue-producing assets reviewed in this study, 86 (21%) were created by the royalty companies and 124 (30%) were represented by single-asset purchase transactions. This suggests that acquiring portfolios of pre-existing royalty assets is the dominant form of asset growth for most royalty firms.

Royalty companies are attractive to investors for many reasons. One important consideration is that it is a scalable business because once a royalty or stream has been obtained, it is relatively simple to track and manage. Thus, operating and administrative staff requirements are small relative to the very large balance sheets and market value of these entities, particularly compared to mining firms. Currently, Franco-Nevada Corporation has 40 full-time employees, Wheaton Precious Metals has 41, and Royal Gold has 31. Dividing their 2021 revenue by these personnel numbers provides a revenue per employee metric of $32.5 million, $29 million, and $20 million, respectively. Few other multibillion-dollar companies can boast of such efficient metrics (Table 9).

Table 9 Comparison of mining royalty company revenue/employee to other industries*

Company[†]	Revenue/Employee, $
Large royalty companies	
Franco-Nevada Corporation	32,500,000
Wheaton Precious Metals Corp.	29,300,000
Royal Gold	19,900,000
Large mining companies	
Glencore	1,609,000
Rio Tinto	1,296,000
Newmont Corporation	846,000
Anglo American	670,000
Barrick Gold Corporation	558,000
Largest tech companies	
Netflix	2,690,000
Apple	2,507,000
Alphabet	1,727,000
Facebook	1,663,000
Amazon.com	297,000

* Adapted from Royal Gold, "Investor Presentation," July 2022, p. 9.

† The large royalty company numbers area as of December 31, 2021. All other company numbers are as of July 2022.

Small management and operation staffs permit company personnel to interact frequently, fostering continuous alignment of everyone's interests with the firm's business strategy and objectives. This engenders effective cost-control measures. Given the organizational structures of royalty companies, tight cost controls lead to higher net cash flow from operations, as demonstrated collectively in Figure 15.

The three companies responsible for the significant jump between 2020 and 2021 are Franco-Nevada ($151.5 million), the IPO launch of Triple Flag Precious Metals ($120 million), and Wheaton Precious Metals ($79.7 million). Their collective cash flows increased by $351.2 million over those of 2020.

This scalability factor positively impacts the earnings of these companies. Figure 16 presents the annual total after-tax earnings for the industry. With such earnings reaching nearly $2 billion, 2014 was a banner year. The launch

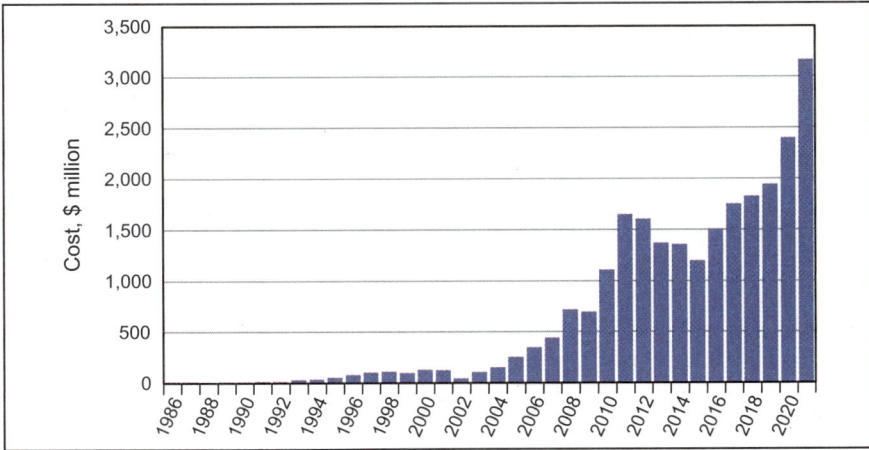

Figure 15 Annual industry cash flows from operations for analyzed royalty companies

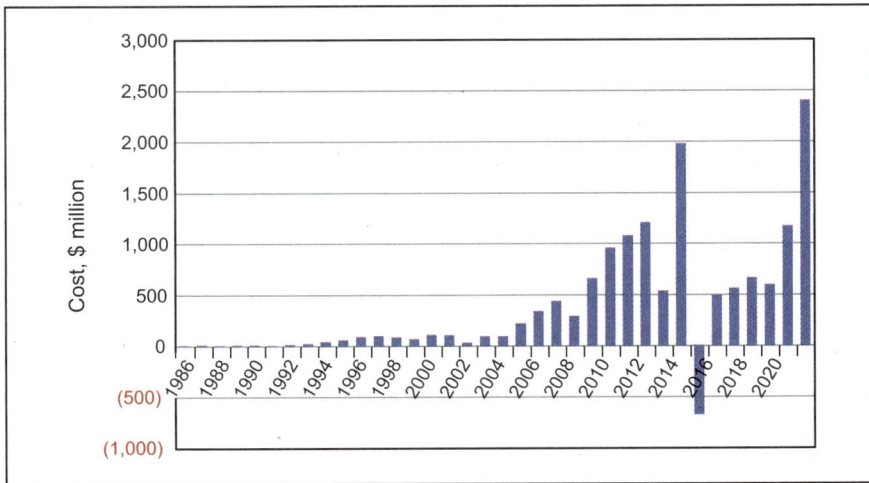

Figure 16 Annual industry after-tax earnings for analyzed royalty companies

of Osisko Gold Royalties in 2015 accounted for 73% of this amount. This was followed in 2015 by Wheaton Precious Metals and Natural Resource Partners both taking large impairment charges, which greatly dragged down the industry totals. However, the rebound was immediate in 2016 and 2017, followed by an almost doubling of 2020 levels over 2019 and increase of another $1.2 billion in 2021.

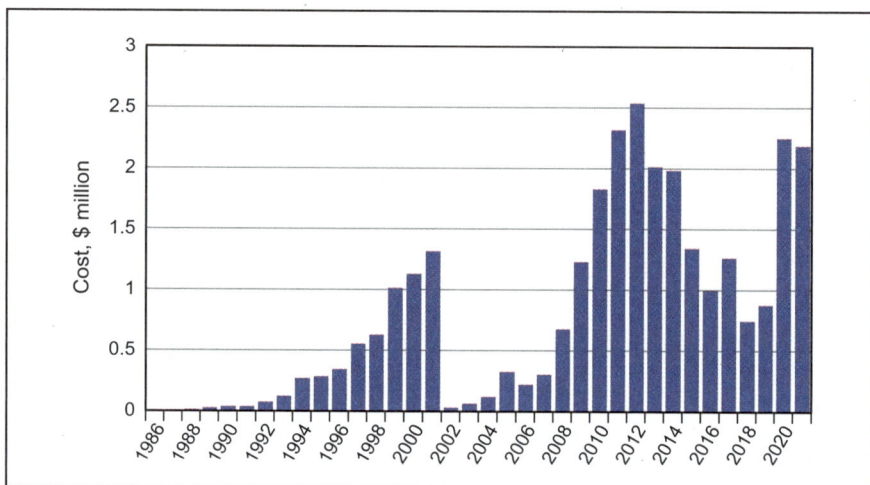

Figure 17 Annual industry working capital for analyzed royalty companies

One of the great benefits of consistently strong earnings is the reliable increase in the firm's working capital. Figure 17 shows the industry's total year-end working-capital levels; these grew steadily from essentially zero in 1986 to $1.3 billion by 2001. As noted previously, Franco-Nevada was acquired by Newmont Corporation in 2002, thereby eliminating numbers from the largest mining royalty company. This resulted in a major step-down in industry working-capital levels over several years in the 2000s. The other companies continued to grow during this interval, and with the re-emergence of an independent Franco-Nevada as Franco-Nevada Corporation in 2007, an era of major increase in working-capital generation returned. This peaked in 2012 with $2.5 billion available across the industry to grow the businesses and increase dividend payouts to shareholders. By 2020, the three largest companies had nearly doubled their working cap levels, achieving levels of more than $2 billion during the latest two years.

These working-capital increases reflect growth in metal commodity prices, the impact of large acquisitions generating increased production, and strong internal organic growth from past transactions stemming from earlier-stage projects reaching maturity.

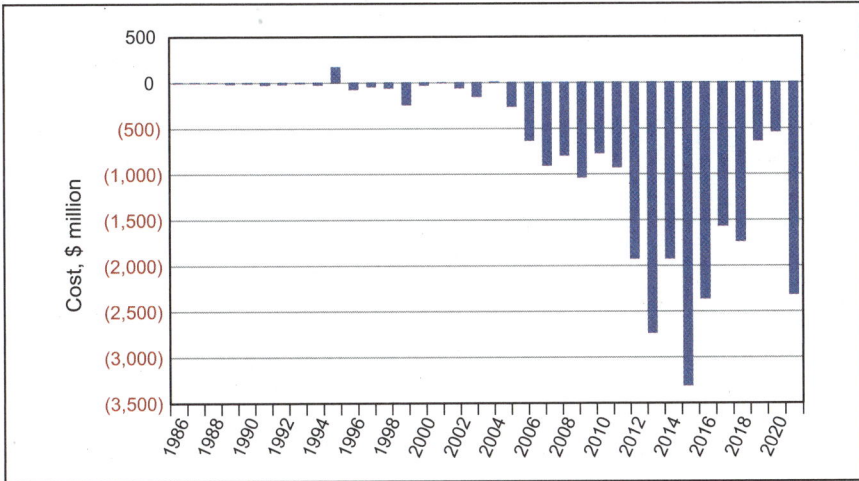

Figure 18 Annual cash flows from investing for analyzed royalty companies

Royalty company strategies use their growing treasure chests to primarily focus on investing in more assets (investing is a cash outflow so the numbers are negative). In the past 8 to 10 years, the sector companies have invested more than $1 billion every year (Figure 18). The present record year was reached in 2015 when $3.3 billion was invested in new assets. As noted earlier, most of these funds are directed to metal stream investments.

Industry participants have a long-established reputation for minimizing the amount of debt carried on their balance sheets, and as their treasuries grew, they reduced outstanding loans as quickly as possible (Figure 19).

This was particularly true until the start of the Chinese supercycle circa 2002. The ensuing increase in commodity prices provided the companies with more working capital that could be leveraged by taking on long-term debt. This activity peaked in 2015 with the industry debt load reaching $3.7 billion. Wheaton Precious Metals was responsible for a large portion of this debt as they financed the creation of several mega-streams; currently, the company has extinguished all of this debt. Franco-Nevada has always had a policy of extreme aversion to debt, while Royal Gold's maximum debt position ($601 million in 2016) has also been eliminated.

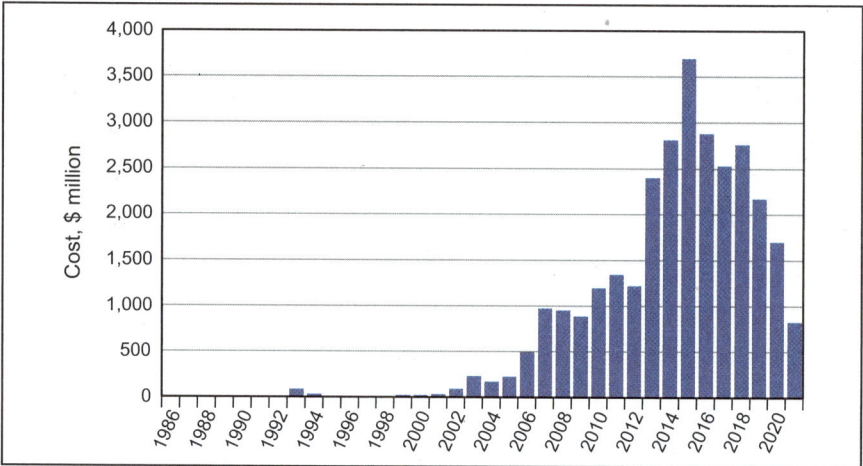

Figure 19 Annual industry long-term debt for analyzed royalty companies

Since 2015, royalty companies have used their large working-capital positions to reduce debt loads, and by 2021, total industry debt had lowered to $821 million. Natural Resource Partners is currently responsible for almost half of this amount. Of the eight largest mining royalty companies, only Osisko Gold Royalties presently has any long-term debt ($90.3 million).

The growth in the royalty industry's working capital has also reduced the need for external financings, as is illustrated in Figure 20. This cash flow is feeding dividend payouts and exercise of stock options, in addition to lowering debt positions.

When companies generate large amounts of working capital, one of their fundamental objectives is to share it to the degree appropriate with their owners—the common shareholders—via regular or extraordinary dividend payouts. This is exactly what the many royalty companies have been doing as shown in Figure 21.

Mesabi Trust is believed to have been the only dividend-paying royalty company in 1986; unfortunately, data for it is not available for the period 1986 to 1996. Franco-Nevada Mining Corporation made its first dividend

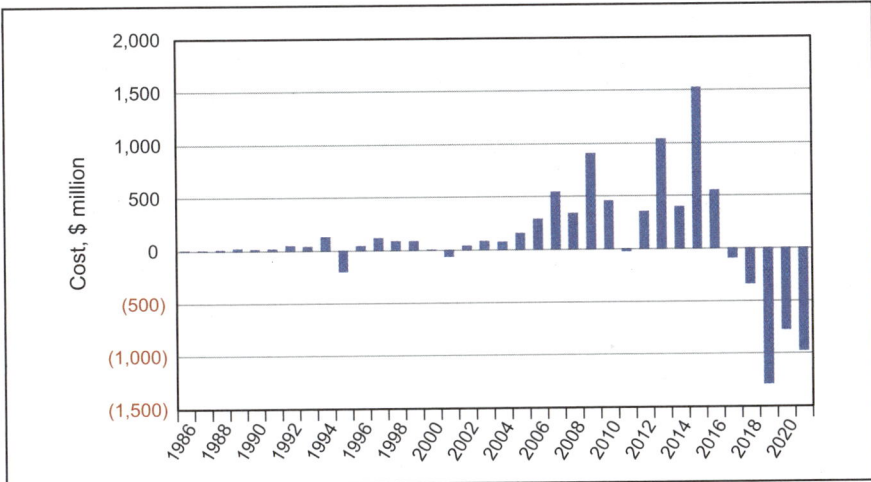

Figure 20 Annual industry cash flow from financings for analyzed royalty companies

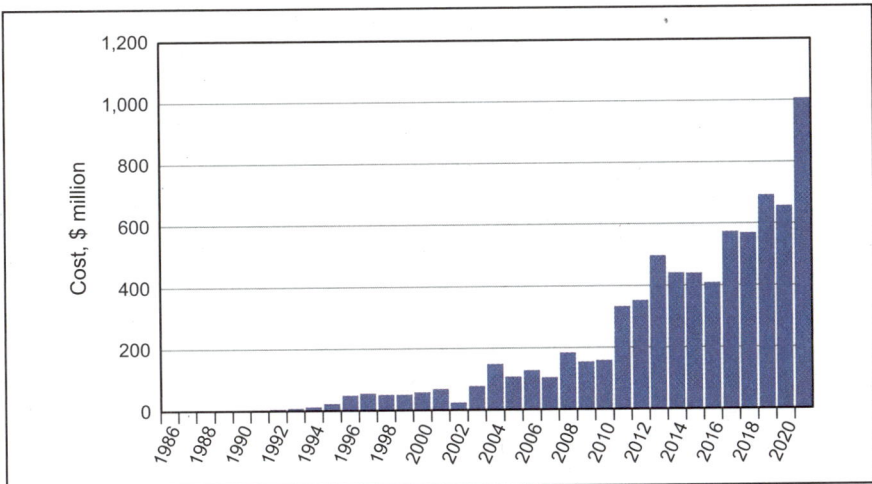

Figure 21 Annual mining royalty industry dividend payments

distribution in 1988; its sister company, Euro-Nevada Mining Corporation, followed in 1991. In 2021, the 15 dividend-paying royalty companies collectively paid out more than $1 billion, representing an industry yield of 1.61%, and many have consistently grown their dividends every year.

INVESTMENT MOTIVES

Investors are attracted to mineral royalty companies for the following reasons:

- One-stop shopping. Each of the large royalty companies has assembled portfolios of several hundred royalties involving a range of mineral commodities, locations, and project status. This provides the opportunity for positive exploration and production surprises, organic internal growth without the requirement of additional investments, and broad portfolio diversification. No mining company can offer similar high upside, low downside economic opportunities across such a wide range of assets.

- The transition of the industry toward the metal stream model provides mechanisms for large step-growth in revenues while also creating investment-instrument diversification.

- Exploration properties are high risk and expensive to explore. Therefore, the value of an exploration royalty is low and remains so until a discovery is made and a mine is built. Royalty companies do not have to fund the high-risk capital investment for mine development and construction but obtain a low-cost entry position to participate in potential future revenues.

- Once a royalty is attached to a prospective mineral lands package, the royalty applies to all future discoveries and resource expansions on the property essentially forever (unless otherwise limited by the negotiated terms of the royalty agreement).

- Every royalty company makes a detailed assessment of the exploration potential of their targets. Franco-Nevada's Goldstrike (Nevada, United States) economic experience epitomizes what happens when a small mine becomes a large mine.

- If the mine expands, the royalty holder benefits from the additional production without having to contribute to the capital costs necessary for such expansion.

- Royalty companies are not mine operators and hence aren't burdened with the high overhead costs and potential cost overruns often experienced by miners via capital or operating cost inflation. In turn, the cash flows can be applied to building the core business of the firm, rather than fixing the operational problems of the underlying enterprise. This makes the royalty model exceptionally scalable.

- Mineral projects frequently pass through multiple owners over the course of their prospective and development lives. However, a properly registered royalty stays as an attachment with the property, regardless of the contemporary operator/owner.

- Similarly, most mineral royalties continue as an obligation across a mine's entire life span and apply as well to new economic resources developed within the subject boundaries. The only way to terminate a royalty is through abandonment of the mineral tenement, the royalty holder opting to sell or terminate the royalty, or the royalty having a cap in monetary or production terms.

- The largest royalty companies are now serving the role as major financiers for mine development. Their financial capacity and present reluctance of conventional funding sources, such as many banks, to support the mineral development provides opportunities to cherry-pick the best projects in need of investment capital.

- Once a royalty, stream, or offtake has been purchased/created, managing it is fairly simple and does not require a large management or administrative staff. This minimizes overhead costs, creates strong profit margins, and makes the business easily scalable.

- Portfolios of producing or near-operation royalties involving a range of commodities provide great upside exposure to metal prices.

- Streams turbocharge this attractive attribute. Once a stream has been created, its ongoing payments tend to be a small fraction of the spot metal price. This provides more leverage in a rising price market.

- ETFs only have exposure to metal prices. Royalty companies similarly participate in increasing metal prices, but also in new discoveries, mine expansions, and improvements in metal recovery.

- Historically, the precious metals royalty companies typically outperform the spot price of gold.

- The royalty company sector is a durable business with staying power because of the strong balance sheets of participants, very high revenue margins, low overheads, and low loss ratios.
- Mature royalty companies have a solid history of returning capital to shareholders.

The mining royalty industry's highly positive financial performance has been rewarded by its high investor metrics. Table 10 compares the upper three tiers of mining royalty companies to four categories of gold producers, three classes of diversified majors and base metal companies, along with iron ore, industrial minerals, fertilizer, and diamond companies. The royalty companies outperform the other public company groupings in share price to net asset

Table 10 Royalty company investment metrics relative to other mining industries*

Company Type	Share Price/ NAV†	Price/ Cash Flow	Price/ Earnings	EV/ EBITDA‡
Tier 1 royalty	2.11×	22.6×	31.5×	21.2×
Tier 2 royalty	1.03×	14.2×	32.4×	13.2×
Tier 3 royalty	1.10×	13.4×	16.9×	20.0×
Major gold producers	1.28×	8.2×	17.5×	6.5×
Senior gold producers	1.08×	6.6×	11.5×	4.4×
Intermediate gold producers	0.77×	5.9×	12.6×	4.3×
Junior gold producers	0.63×	5.2×	7.9×	4.3×
Diversified majors	0.91×	4.9×	7.2×	3.3×
Senior base metal producers	1.18×	7.3×	10.7×	5.5×
Intermediate base metal producers	0.76×	4.3×	13.9×	7.8×
Iron ore producers	1.80×	6.8×	8.9×	4.1×
Industrial mineral producers	N/A	6.5×	13.5×	8.1×
Fertilizer producers	N/A	9.0×	10.3×	7.0×
Diamond producers	N/A	2.8×	7.7×	3.4×

* Data from CIBC, December 31, 2021.

† NAV = net asset value.

‡ EV/EBITDA = enterprise value (EV) to earnings before interest, taxes, depreciation, and amortization (EBITDA).

value (NAV), price/cash flow, price to earnings, and enterprise value (EV) to earnings before interest, taxes, depreciation, and amortization (EBITDA).

These superior financial metrics reflect the innovative deal structures the royalty companies have entered into along with the sector's unique ability to minimize costs because of the scalability of the business. This has provided these companies with impressive cash flows that permit ever greater investments, leading in turn to larger revenues and earnings, and the ability to increase their dividend payouts or conduct share buybacks.

EVOLUTION OF AN INDUSTRY

In its nascent period from 1986 to 2000, before the number of royalty companies had not yet totaled 10, the Franco-Nevada group of companies (Franco-Nevada Mining Corporation, Euro-Nevada Mining Corporation, and Redstone Resources) had easy domination of the mining royalty industry. Once they got their ball rolling, they played in an open field of opportunities. Royal Gold came to the party in 1991 and began competing with the Franco-Nevada group in the Nevada (United States) goldfields. While Franco-Nevada's focus was on royalties sited in the Carlin (gold) trend, Royal Gold's focus was on the Cortez/Pipeline (Nevada, United States) area.

From 2002 to 2006, mineral companies were enticed to sell their traditional royalty holdings as packages to the mining royalty companies. This permitted cleanup of balance sheets, reduction of management time directed to non-core activities, and the opportunity to take advantage of an increased metal prices environment.

Royalty creation schemes emerged during 2006 through 2009 as a mechanism for mine operators to raise capital without assuming more debt or shareholder dilution. The concept involved the royalty firms "buying" a newly created royalty on the mine asset for an up-front payment. The problem was that even a large 5% NSR did not generate sufficient future revenues to justify a materially impactful capital infusion for the developing mining operation.

Silver Wheaton, the predecessor company to Wheaton Precious Metals, had a different idea, which addressed the size of capital obstacle. They invented the metal stream purchase (MSP) model. Its advantage was that it could provide a meaningful amount of immediate funding for the mine operator by selling to Wheaton the mine's by-product silver output. Furthermore, because copper mines receive lower price/NAV multiples than precious metals mines, selling off their future by-product precious metal production effectively provided the base metal producer a precious metals multiple on the asset sale. This made streaming arrangements very attractive, especially to base metal companies with material, but not financially essential, precious metals credits.

The credit crisis of 2007 to 2009 greatly inhibited debt financings in all industries, but especially in the mining space. It drove many of the traditional mining banks and other traditional capital sources out of the sector. This further compelled the mine operators to use streaming financings.

By 2010, most of the world's largest mining royalties had been acquired by the royalty companies. The one exception was Iluka Resources' giant royalty on BHP Billiton's Mining Area C iron operation in Western Australia. Iluka rebuked many offers over the years but finally spun out this royalty, along with others, into Deterra Royalties in 2020.

With industry financial growth and a reduced number of major acquisition opportunities, the larger royalty companies faced a serious dilemma. Investors expect companies to increase revenues every year and preferably at a rate in excess of 10%. For firms generating revenues greater than $100 million, this means they had to purchase royalties that provided at least $10 million in annual payments. By 2010, however, most of these had been acquired by the competing royalty participants and were no longer purchase candidates. Alternative metal streams quickly evolved as practical strategies to maintain revenue growth objectives. The classic royalty company began transitioning into the streaming arena. Meanwhile, the high price-to-net-asset-value (P/NAV) multiples being realized by the largest royalty players became an incentive for additional start-up firms, often backed by private equity, to join the industry.

To become a serious contender, the new companies needed to differentiate themselves, and they did so in many innovative ways. Uranium Royalty Corporation focuses only on uranium royalties, while Nova Royalty Corporation hunts for copper and nickel royalties. Private company Lithium Royalty Corporation only chases lithium assets. Not surprisingly, none of these new entrants focus on precious metals, likely because of the competitive dominance of the Tier 1 firms in those commodities.

Some new precious metals–focused royalty firms have emerged but as spin-offs from existing mineral companies. Maverix Metals came out of Pan American Silver Corporation, Great Bear Royalties Corporation was spun out of Great Bear Resources, and Osisko Gold Royalties is a member of the Osisko Group of Companies.

Another block of companies began as exploration outfits and converted to a royalty format once they accumulated a sufficient number of royalties to justify the strategy revision. This group includes Altius Minerals Corporation, Altus Strategies, Ely Gold Royalties, EMX Royalty Corporation, High Desert Mineral Resources, Mundoro Capital, Premier Royalties, Terraco Gold Corporation, Vaaldiam Mining, and Xtierra.

The downturn in the markets circa 2009 saw the exit of traditional mining banks and created a large gap in mine financing options. The private equity world stepped in to provide the capital for new and expanded mines. From a dozen or so fully dedicated mining private equity firms, Triple Flag Precious Metals emerged with an IPO in mid-2021 and instantly became one of the largest mining royalty firms.

As mentioned before, most of the existing royalties are connected to early stage projects. Investors are notoriously impatient, however, and will not wait years for these projects to be placed into production and provide a return on their investments. This forces the royalty companies to pursue near-term production opportunities at the expense of investing in longer-term projects that may have potentially much higher returns and longer mining lives.

There are several estimates on the number of mining royalties that currently exist worldwide. Most estimates are between 5,000 and 8,500, suggesting that there is still plenty of product left to be acquired by knowledgeable players.[3]

THE INDUSTRY'S FUTURE

The three major royalty companies are extremely well positioned to continue dominating the industry for the foreseeable future. With their large financial treasuries, access to low-cost capital, superior share-price performance, and market capitalization size, they enjoy the benefits of strong support by institutional investors. Their financial firepower allows these companies to close multi-hundred-million-to-billion-dollar deals, transactions firms in the other tiers can struggle to take on. Most of anticipated future growth will come from execution of metal streaming transactions connected to development or producing projects. They will also experience substantial organic growth as their investments in earlier-stage assets mature.

The *Engineering and Mining Journal* estimated that there are more than $1 trillion worth of mining projects in the development stage.[4] These constitute a fertile hunting ground for the royalty companies, whether they are contributing construction capital or financing acquisitions.

Additionally, some of these royalty companies are beginning to take equity positions in mining companies, behaving more like merchant banks. They are highly sensitized to the lower P/NAV multiples of holding companies, so they will only expand in these arenas on a small or judicious basis.

Furthermore, because the bigger companies have large existing and projected revenue streams from their multiple assets, the market will permit them to purchase assets with longer timelines to production: something the smaller companies logically are not capable of or should avoid.

---

3. Data from Vox Royalties to Douglas Silver, personal communication, October 6, 2022.

4. Joe Govreau, "Mining Industry Embraces Decarbonization Sea Change in 2022," *Engineering and Mining Journal* (January 2022), 19.

The Tier 2 companies will continue to fight for that coveted fourth place in the market capitalization ranking. This will be achieved by buying smart, increasing commodity prices, accelerated organic growth through the consolidation of individual Tier 2 through Tier 4 companies, and, of course, good old-fashion luck.

An excellent example of such advancement tactics was seen in 2022 with Sandstorm Gold's simultaneous purchases of public Nomad Royalty Company and private BaseCore Metals. This combined $974 million acquisition deal should well position Sandstorm for increase in its market capitalization.

Another recent example involves Gold Royalty Corporation. During 2021, the company acquired three public royalty companies: Ely Gold Royalties, Abitibi Royalties, and Golden Valley Mines and Royalties. These forms of industry cannibalism will most certainly continue, whittling down the number of mid-tier players, while simultaneously providing market space for the smaller companies to grow.

Opportunities exist for new royalty/streaming models to evolve. There is little question that the invention of the metal stream model revolutionized the mining royalty industry and the financing of mine capital. It provided a unique way for investors to obtain large revenue streams while also mitigating the increasing funding challenges plaguing the mining industry with conventional financing sources. Over the past two decades, both royalties and streams have been adapted as sector participants increase in number, and innovation can represent the difference between winning and losing a deal. These instruments are now a key source of funding for mineral development, and in light of the major expansion of mineral extraction required in coming decades to underpin global energy transition, they will play an ever-increasing role.

The metal streaming model has now been around for 17 years and one has to wonder whether it is time for a new financial instrument that better aligns the interests of both the miner and the royalty companies. From the 1980s, royalties were essentially ignored by the mining companies. Today, royalties and streams form an integral component to many mine financings, and the mining companies have developed a sophisticated understanding of royalty and streaming structures. This naturally leads one to believe that both parties will invent a new structure that more closely aligns both parties' interests, while introducing investors to a new generation of value creation opportunities via the mining royalty industry.

2

FRANCO-NEVADA

TSX | NYSE: FNV

franco-nevada.com

Pierre Lassonde and David Harquail

PART 1: FRANCO-NEVADA MINING CORPORATION LIMITED

Pierre Lassonde

This is the story of the original Franco-Nevada Mining Corporation Limited (FN) that Seymour Schulich and I started in 1982 and ran up to the merger with Newmont Mining Corporation in 2002. We hope it reflects the excitement and joy we had inventing the public mining royalty business and creating a lot of wealth for our investors.

BIRTH OF THE MINING ROYALTY BUSINESS (1982-1987)

In the early 1980s, both Schulich and I were money managers at the investment firm Beutel, Goodman & Company in Toronto. Schulich focused on oil and gas investments, and I managed mining investments, including a number of gold funds. We both loved Nevada—Schulich for the poker and me for the skiing—and we wanted a way to write off our trips as business expenses. We formed FN in 1982 and then took it public on the Toronto Stock Exchange (TSX) in September 1983, raising Can$2.3 million at Can$0.35 per share. FN started life drilling exploration prospects in California and Nevada.

It didn't take long to realize that drilling holes without any geological expertise was the fastest way to the poorhouse. If FN was going to endure, it needed a source of cash flow. It was Schulich who asked, "Aren't there any royalties in the mining business like in the oil business?" That idea led us to answer a box ad in the *Reno Gazette* for a royalty for sale. In March 1986, we paid $2 million for a royalty on the Goldstrike mine in Nevada (United States). At the time, it was a small heap-leach mine on the Carlin trend operated by Western States Mining, producing roughly 44,000 ounces per year. What we liked about the royalty was that (1) we would get our money back in five to seven years, faster if the gold price picked up, giving some gold price optionality; and (2) several drill holes had intersected high-grade mineralization at depth. There was exploration optionality!

Shortly after FN bought the royalty, Barrick Gold Corporation (American Barrick at that time) purchased the Goldstrike mine for the then princely sum of $62 million. The consensus view was that the high-grade intercepts in the area had been down veins with no lateral extent or depth potential. Bob Smith, who was president of Barrick, recognized the potential and immediately on acquisition had Barrick follow up the previous results. The rest is history. Ultimately, more than 50 million ounces would be proven up at Goldstrike. To date, our $2 million royalty investment has generated close to $1 billion. This was our first royalty purchase: It was as if we had started playing the game of golf and got a hole-in-one with our first swing.

FN was now in the public royalty business. In our fiscal 1987 annual report (our fiscal year end was March 31), we wrote that

> We have found a niche in the gold mining business that few others are interested in but can be quite profitable for a small, aggressive and shareholder-oriented company. …The key points of our philosophy are:
>
> 1. We like royalties…and will be exclusively working on the purchase of royalties. …
>
> 2. We are not operators. We do not want to administer a big staff. …
>
> 3. It is our intent to continue to maximize the value of FN shares.

EARLY YEARS AND EURO-NEVADA MINING CORPORATION

In the early years of the royalty business, we continued to work from the Toronto offices of Beutel, Goodman & Company, and FN was operated on a part-time basis similar to a closed-end fund. David Harquail joined in 1987 to add business development support. Subsequently, Ron Binns joined to take care of the financial reporting and act as chief financial officer (CFO). Sharon Dowdall was our outside legal counsel and would eventually join as our chief legal officer (CLO). In the United States, our outside legal counsel was Craig Haase. He also joined us, and his Reno law office became the home for FN's geologists and landmen. A key early hire was Steve Aaker who would become our most trusted geological adviser. In those early years, we had little competition in buying royalties. Operating companies were focused on managing yellow trucks and saw royalties simply as a cost of doing business. We made it our business to identify prospective properties, go to local courthouses to search underlying ownership, and introduce ourselves to the royalty owners. We would sometimes visit royalty owners and get chased away with a shotgun. Other times, we would buy royalties from their heirs or vindictive divorcees. FN was known as the "odd lotters" of the mining business—buying what no one else wanted.

As Goldstrike rapidly grew, Schulich and I realized that FN had very little tax basis and would soon be paying substantial taxes. As money managers, it made financial sense for us to sell Goldstrike to an operating company that could better shield the royalty income. We created Euro-Nevada Mining Corporation Limited (EN) in honor of our European investors who had supported the early days of FN. The EN initial public offering (IPO) was done as a dividend spinout to FN shareholders. The plan was for FN to be sold holding only the Goldstrike royalty and EN would continue with all the other royalties acquired to that date. The timing for the EN IPO and Nevada investor tour was certainly memorable. It was the same fateful week of the Black Monday stock market crash—October 1987.

This would have been a very different story if we had sold Goldstrike as planned. Peter Munk, who was chairman of Barrick, looked at the opportunity and noted that analysts were valuing Barrick on 100% of the ounces it produced or had in reserves and that none of them was deducting the royalty ounces. Barrick passed as buying the royalty would not add to its market valuation. Regardless, the largest influence in us not selling was Brian Meikle, Barrick's chief geologist. Meikle was a soft-spoken man of few words, but when he spoke, you listened. When I told him our plans, he said, "Pierre, Goldstrike is not the discovery of a lifetime, it's the discovery of three life-times!" The next day, Schulich and I canceled our plans to sell FN. Great ore bodies tend to keep on giving. Owning an interest in a great long-term ore body takes precedence over short-term financial engineering.

We would manage FN and EN as separate royalty companies for the next 12 years. Originally, FN was positioned as the value company with a 50% dividend payout policy while EN was the growth royalty company. FN started paying its first dividends in 1988, and EN followed in 1991. Over time, the two companies increasingly undertook joint ventures (JVs) on the same assets. In September 1999, the two companies were merged to improve the appeal to large institutional investors. (To simplify the ongoing story, references to FN will treat both FN and EN as having always stayed as one company.)

GROWING THE ROYALTY PORTFOLIO

Our royalty acquisition focus started on the more prolific gold camps in the world, including the Carlin and Getchell trends in Nevada (United States), Hemlo and Holloway in Ontario (Canada), and the Kalgoorlie belt in Western Australia. Royalties purchased in Nevada included Dee, Hollister, Getchell, Maggie Creek, Rosebud, and Santa Fe. Just across the border in California, we bought royalties on Castle Mountain and Briggs. In Canada, we added royalties at Eskay Creek (British Columbia), Mouska (Québec), Harker-Holloway (Ontario), Kirkland Lake (Ontario), and Goldfields (Saskatchewan). Internationally, we added royalties in Australia at Henty (Tasmania), Browns Creek (New South Wales), and New Celebration (Western Australia) as well as Mount Muro in Indonesia and Mara Rosa in Brazil. We also expanded into platinum group metals with the purchase of royalties at Pandora in South Africa and Stillwater in Montana (United States) ($36 million in March 1998).

A royalty creation innovation that we developed was to sponsor new exploration companies in return for a royalty on their projects. In the pre-Bre-X Minerals days of the early 1990s, it was easier to raise risk capital for international exploration. One company that FN sponsored was Craig Nelsen's Metallica Resources, which had properties in the KM 88 region of Venezuela. To Nelsen's credit, he eventually pivoted to Mexico where we ultimately ended up with a royalty on the Cerro San Pedro mine. Another company was Franco-Or that focused on French Guiana. In Cuba, we sponsored Caribgold, which was led by David Bell, the geologist who discovered Hemlo.

I was part of an investor tour to Cuba and had the honor of talking to Fidel Castro who had read my book on gold. It was suggested to me that I present Castro with one of our FN-labeled gold pens. Castro's security team did not take kindly to it: Little did I know that the CIA had made past assassination attempts with poison pens!

Another FN initiative was to buy strategic stakes (up to 20%) in companies with good assets that we wouldn't have to run ourselves. In the latter half of the 1990s, we bought significant positions in Echo Bay Mines, Aber Diamond Corporation, Inco's Voisey's Bay (Labrador, Canada), and the San Juan Royalty Trust. Our experience is that with a significant strategic stake, we could both influence strategy and be a catalyst. We were able to realize significant gains in all those positions.

COMMODITY DIVERSIFICATION

In 1989, we set up David Harquail to run a separate public company, Redstone Resources, to focus on non-precious minerals. It bought interests in Falcondo nickel (Dominican Republic), Robinson copper (Nevada, United States), Midwest uranium (Saskatchewan, Canada), and another 20 mostly preproduction royalties including the Rosemont copper deposit in Arizona (United States). Redstone was profitable, but growth was slow. Royalties in this segment tend to be large in scale, infrequently available, and costly to acquire. The base metals industry has a fraction of the operators that exist in the gold business, and the time line to development is multiple times longer.

We are still waiting on Rosemont's permits. In late 1995, Redstone's royalty portfolio was rolled into FN as its metals division. We continued to acquire non-precious mining royalties as FN's financial strength and patient investment approach make it the best home for these assets.

In contrast, the oil and gas industry is a very rich and active royalty space with many operators, millions of available fractional royalty interests, and rapid development time lines. This is what had inspired Schulich to suggest FN should look for royalties. In 1992, FN began adding oil and gas royalties in western Canada to its portfolio. We were cognizant of the premium multiples the market places on precious metals and appreciated the concerns of gold analysts and gold fund managers. At the same time, our growing generalist institutional investor base liked the increased portfolio diversity and broader investment opportunities. We struck a balance by committing in bold letters in our annual reports to not allow our metal and oil and gas royalty assets to contribute to over 25% of FN's earnings base. This was comparable to the non-precious earnings then being realized by many of the large gold operating companies. FN has been well served through the cycles by having a diversified commodity portfolio.

ROYALTY CREATION

FN had started life as an exploration company that then discovered the royalty business. But doing exploration was also another way to create new royalties with the added benefit of establishing a larger active presence in the United States for tax purposes. Our landmen and geologists were tasked to assemble land packages on promising targets, do enough drilling to add geological value, and then vend the package to an operator in return for a royalty. Our initial focus was on land packages along the Carlin trend. The team was successful in optioning various packages in return for royalties, but we knew the odds were very long in exploration. In the end, these royalty creation efforts vastly exceeded all our expectations. (The Midas discovery is discussed in more detail in the next section.)

In Canada, our strategy was to focus on the great mine trends. In 1990, FN paid Can$1.6 million to acquire full ownership of 620 acres covering the western downdip extension of the great Hemlo ore body in Ontario. Teck Resources Limited and Homestake Mining Company were partners in the neighboring Page-Williams mine (Ontario) but discounted the downdip potential as being unproven and too far in the future to pay that much. We felt the operators' perspective would change over time as mining moved closer to the boundary. And to help their perspective, we later invested Can$7 million doing three years of underground drifting and drilling to outline a 1.9 million ounces resource. We eventually converted our claims into a 50% net proceeds interest (NPI) and 3% net smelter return (NSR) on those ounces. In 2020 alone, these royalties generated to the new FN almost $70 million in revenue. The lesson here is that operating management teams often limit capital decisions to shorter discounted value measures and undervalue the exploration optionality.

FN didn't have the same tax motivation to hire its own geological teams in Canada. Instead, our approach was to JV with junior explorers on great land positions and later convert our interest into a royalty. An example was our JV with Queenston Mining that ultimately was converted into royalties covering most of the Kirkland Lake camp. Our biggest Canadian success was our JV with Pelangio Resources (now called *Pelangio Exploration*) on the Detour Lake property (Ontario). Our JV cost was Can$2 million, which was eventually converted into a 2% NSR. In 2021 alone, that royalty generated more than $25 million to the new FN, and the mine is being expanded. With over 20 million ounces in reserves today, we expect Detour to ultimately generate more than $1 billion in royalties for the new FN—another example of exploration optionality.

MIDAS/KEN SNYDER MINE

By 1994, our royalty creation efforts in Nevada had extended to the northern end of the Carlin trend where there was an historic mining camp known as *Midas*. Our chief geologist, Ken "Dry Hole" Snyder drilled a target known as *Rex Grande*. Despite Snyder's nickname, his drilling led to a massive high-grade epithermal gold and silver vein called *Colorado Grande*. FN unexpectedly had a discovery on its hands that would ultimately produce 2.2 million ounces of gold and 27 million ounces of silver. The resulting appreciation in FN's market value on the news of the discovery made it impossible to then convert Midas into a royalty of equivalent value. Till then, we had been called the "Gucci shoe miners," as our business model was to never build or run any mines. Now we would need to develop Midas ourselves. In our 1997 annual report, we wrote:

> Life is a poker game. A person or company must play the cards they are dealt. Midas represents a departure from the strict royalty role we have followed to date. We outline herein quite clearly why we believe Midas has the characteristics that make it amenable to a contract mining approach. …We want to assure our shareholders that we remain committed to the concept of royalty creation.

Andre Douchane was hired to build Midas and supervise the mining contractor. Douchane successfully built the underground mine along with a 1,200 ton-per-day mill under budget for $84 million. Commercial production began ahead of schedule in late January 1999, and Midas was a profitable asset for FN despite the gold price averaging only $275 per ounce from 1999 to 2001. In May 2001, FN not only finally achieved its original goal of converting Midas into an important royalty, but it also obtained as part of the consideration a 19.9% strategic stake in Australia's largest gold company, Normandy Mining. In just a few quarters, that strategic stake would ultimately be worth a net $460 million to FN. This was the absolute bottom of the gold market and yet our royalty creation efforts had created substantial value for FN's shareholders.

NEWMONT MERGER

From 1999 to 2001, gold averaged only $275 per ounce, and it was killing the gold industry. Apart from Midas, few mines were being built, the risk money had disappeared, and the junior explorers were in hibernation. Royalties work best when there is risk capital available to spend on the properties and that was not happening. Schulich and I were convinced we were close to the bottom in gold. Why not look for ways to get more torque to gold by merging with a large unhedged gold producer? FN then had a market cap of about $2 billion that ranked it among the five most valuable gold companies in the world. In June 2000, we announced an agreement to merge with Gold Fields Limited, which had the largest unhedged gold reserves in the world. However, three months later, the South African government blocked the merger. It was ironic that the next year, we unexpectedly found ourselves having to deal with another South African company. In the summer of 2001, AngloGold announced a bid for Normandy Mining without even checking with Normandy's major shareholder. FN had a 19.9% position in Normandy as a result of selling Midas just a few months earlier. FN suddenly was in a position to play kingmaker.

One option was to make a deal with Barrick. Our companies had a long history together and were both based in Toronto. Barrick had recently purchased Homestake Mining and was the number one gold company in the world. However, Barrick was the mostly heavily hedged of the gold companies. If Munk were willing to stop gold hedging, we could help cement Barrick's position as number one in the industry. Munk was unwilling, and that hedge position ultimately cost Barrick more than $5 billion in losses.

In contrast, Wayne Murdy, the incoming CEO of Newmont, was much more receptive to ending hedging. The combination was ideal with FN having a strong balance sheet, Newmont having debt but a strong development team, and Normandy having a strong pipeline of projects. Our key negotiations were set to take place on September 11, 2001. Yes, 9/11. Needless to say, the Newmont team never left Denver that day.

Ultimately, we did get together and made it work. And despite four rounds of bids by AngloGold, our three-way merger with Newmont and Normandy was consummated in February 2002. On the final day of trading for FN, it was valued at close to $3 billion, and the combined new Newmont was worth almost $10 billion, ranking it as the largest and most valuable gold company in the world.

FRANCO-NEVADA MINING CORPORATION LIMITED POSTSCRIPT

Both Schulich and I take great satisfaction in looking back at what was accomplished with the original FN over almost 20 years:

- We created a successful new business model for the gold industry that others would want to emulate. Today, we count 46 public mining companies that now call themselves royalty companies.
- We achieved a $3 billion outcome for our investors against less than $1 billion in shareholder equity, demonstrating real value creation.
- We paid more than $250 million in dividends over 14 continuous years of dividend payments.
- All of this was achieved despite a bear market for gold that ended only in 2002.

When we sold the original FN, we were right in our call that the gold price could only go up. What we were wrong about is that we expected there would be more torque with the gold operating companies. The reality turned out to be that the operating companies allowed costs to rise along with the gold price. Margins stayed the same despite a rising gold price. Ultimately, we would have to get back to the royalty business, which leads us into the second part of the story.

PART 2: FRANCO-NEVADA CORPORATION

David Harquail

In the preceding "Part 1," Pierre Lassonde covered the story of the original FN that he and Schulich led from 1982 until it was merged into Newmont Mining Corporation in 2002. This is the story of the next generation royalty company that was spun out of Newmont in 2007 as Franco-Nevada

Corporation (FNV). The success of the FNV story reflects the DNA it inherited from the original FN.

NEWMONT ERA (2002-2007)

The $10 billion merger of the original FN with Newmont and Normandy Mining was completed in February 2002. Although the FN staff was small in number at only 20 employees, some large roles were entrusted to the FN team. At the Denver head office, Lassonde took on the role as Newmont's president. Newmont Capital was established as its merchant banking arm and charged with managing overall business development and the royalty portfolio. I relocated from Toronto to be Newmont Capital's managing director. Aaker relocated from Reno to support geological evaluations. The former FN Toronto head office effectively became a branch office of Newmont Capital and included key FN employees such as Dowdall who headed legal and Geoff Waterman who managed the energy assets.

The post-merger focus for Newmont was to reduce debt. Former FN assets, such as its holdings in Echo Bay Mines, Aber Diamond, Inco, and direct property interests at Hemlo, Detour, and Kirkland Lake, were sold or converted into royalties. The same was done with many non-core Newmont and Normandy assets. The exploration assets being sold had geological potential that had attracted the interest of major gold company geologists but nevertheless were deemed non-core because of the richness of the newly combined portfolio. When Newmont Capital was first created in 2002, it inherited 60 mining royalties from the original FN. By 2007, the combination of existing royalty portfolios from the three companies along with the creation of royalties on sold assets expanded the total mining royalty portfolio to 290 mining assets (21 operating, 15 advanced, and 154 exploration projects). During the Newmont era, the focus on debt reduction rather than buying royalties left the door open for Royal Gold, Silver Wheaton Corp., and International Royalty Corporation to grow their portfolios through acquisitions during that time. However, the exploration optionality gained from the newly expanded royalty portfolio would prove even more fortuitous for the future FNV.

Franco-Nevada Corporation Spinout (2007)

By 2007, Newmont had a new CEO and the focus was on adding to the gold projects pipeline without equity dilution. At the same time, Newmont Capital was enjoying windfall profits from its energy royalties in Western Canada because of record high oil prices. Newmont saw an opportunity to swap non-core royalty and energy assets for gold assets, and in the summer of 2007, investment bankers were invited to present how Newmont could best monetize the royalty assets. Royal Gold, which was a leading royalty company also based in Denver, had already expressed a strong interest, and Newmont's U.S. bankers saw a trade sale to Royal Gold as the most likely course. Canadian investment bankers saw the positive investment legacy left behind by the original FN as an opportunity to realize a better valuation through an IPO of a reborn Franco-Nevada.

A competitive dual-track process was launched in the fall of 2007 with the U.S. bankers leading the trade sale track. Leading the IPO track were BMO Financial Group and the Canadian arm of UBS Canada. With Newmont Capital disappearing, it was an easy decision for Denver-based Lassonde, Aaker, and me to elect to be part of the FNV IPO team. The former FN Toronto office was still mostly in place including Dowdall as CLO and Waterman to manage the energy royalties. Also joining the FNV team was Paul Brink, who had joined the Toronto office in 2006 to be part of business development.

Over the previous six years since FN's merger with Newmont, there had been significant growth in the public company royalty sector. In November 2007, a snapshot of the space looked like Table 1.

A target enterprise valuation for FNV was estimated at $1 to $1.2 billion, somewhat above Royal Gold's but a fraction of Silver Wheaton's. This reflected that only 50% of FNV pro forma revenues were from precious metals because of the then strong revenues being achieved from record oil prices. Also, the original FN's large cash treasury, strategic equity stakes, and the Midas royalty would not be part of the newly incorporated FNV. The new company would start with some debt and an untested management team.

Table 1. 2007 Industry pro formas*

Pro Forma	Silver Wheaton	Royal Gold	International Royalty	Franco-Nevada
Market cap	$3,600	$875	$450	—
Net debt	$430	($55)	($40)	—
Enterprise value	$4,030	$820	$410	—
Revenue	$171	$48	$42	$94
EBITDA[†]	$111	$37	$32	$84
Margin	65%	76%	77%	89%
Number of royalties/streams	5	39	65	290

* Figures are shown in U.S. million dollars.

† EBITDA = earnings before interest, taxes, depreciation, and amortization.

The dual track for the Newmont Capital royalty portfolio was competitive between Royal Gold's bids and the FNV IPO. Royal Gold may have been limited by having Merrill Lynch as its adviser. In late 2007, Merrill was at the heart of the then burgeoning subprime crisis. In contrast, a pre-IPO tour of North American and European institutions had shown strong interest in FNV. In late November 2007, the button was pressed on the formal IPO process, and marketing started in Europe. The reception was beyond our expectations, and before we started our North American marketing, there were already enough orders to ensure a successful IPO.

A highlight for Lassonde and myself was when we visited Fidelity Investments in Boston. They had been hugely successful investors in the original FN, and we knew they would be interested in the new FNV. Will Danoff, the incredibly successful manager of the more than $100 billion Contrafund, was there with over a dozen other Fidelity managers. However, Fidelity also has the reputation for placing their orders late and for trying to shave the price. Lassonde started the meeting by saying that we were there simply as a courtesy, and that we didn't need Fidelity for this deal. If Fidelity tried to shave the price or was late with their order, it was out of the deal. The Fidelity fund managers were in shock and turned to Danoff to see his reaction. In that moment, you could have cut the air with a knife. When Danoff started laughing, they all howled in laughter. Fidelity put in a $100 million order without an argument on price and well before the deadline. Fidelity has been FNV's largest shareholder since its launch and, after more than 14 years, remains so today. We couldn't ask for a more supportive shareholder.

Ironically, after all the worries about whether the largest mining IPO in the TSX history could be pulled off, Lassonde and I then had the opposite worry about being too successful! As the IPO gathered momentum, the projected IPO pricing was moved progressively higher. It meant more proceeds to Newmont, but a higher IPO price would make it more challenging for us to deliver returns to the new FNV shareholders. In the final days, we had to threaten to walk if the IPO price was set even higher. In the end, the FNV IPO share price was finalized at Can$15.20 (in late 2007, the Canadian and U.S. dollar were about equal), and FNV began trading on the TSX on December 20, 2007. With the greenshoe option, FNV raised a record $1.3 billion for a mining IPO and paid Newmont $1.2 billion for the royalty portfolio. FNV was able to start 2008 as a new company fully independent and debt free. Interestingly, a significant number of Canadian institutions and analysts regarded FNV at that price as overvalued and missed out on what turned out to be an outstanding investment opportunity.

Was $1.2 billion a good price for the royalty portfolio? One simple measure is looking at just the IPO assets from the end of 2007 to the end of 2020. In those 13 years, the IPO portfolio has paid out over $1.6 billion. Yet the ounces associated with those same properties has tripled over that time at no cost to FNV. Key contributors include Detour Lake, Tasiast (Mauritania), and Duketon. We continue to be amazed at the power of the royalty business model and the exploration optionality of being exposed to great geology. Our IPO fears of overpaying for the portfolio were quickly replaced with the recognition that we got a bargain.

FRANCO-NEVADA CORPORATION (2008 TO PRESENT)

In January 2008, the new FNV board gathered and set the company's strategic priorities:

1. Grow the precious metals royalty revenues.

2. Become the leader in the mining royalty space and a "go-to" gold stock.

3. Achieve a $10 billion market cap within 10 years.

We exceeded our own expectations in achieving all these goals quickly. Why did we choose the $10 billion target? It was Newmont's market cap at that time, and it seemed a fitting aspiration for the spinout company. FNV's market cap eventually would exceed $30 billion. Clearly, we weren't ambitious enough.

Brink led business development, kicking off FNV's first year by paying more than $100 million to acquire a royalty on Newmont's Gold Quarry operation in Nevada. The next year again more than $100 million was spent on royalties at Coeur Mining's Palmarejo mine in Mexico and Newmont's Subika mine in Ghana. Supplementing the acquisitions, FNV's large portfolio of pipeline royalty assets demonstrated their ability to contribute organic growth as existing mines were expanded (Tasiast, Stillwater, Marigold, Macassa [Ontario, Canada], etc.) or new mines were developed or restarted (Duketon, Detour, Golden Highway, etc.) all at no cost to FNV. This organic growth contribution has been and continues to be a key contributor to FNV's outperformance. About this time, Sandip Rana joined to become FNV's CFO. He had been a part of the original FN so knew the business and culture and was able to hit the ground running.

As FNV's revenues became more weighted to precious metals, its share price was rerated upward. A step change occurred in 2011 with both a New York Stock Exchange listing and the acquisition of Gold Wheaton Gold Corporation. That acquisition gave FNV the infrastructure to begin its own streaming deals. In 2012, FNV made a $1 billion gold and silver streaming commitment to support the construction of the giant Cobre Panama (Panama) project. That commitment was ultimately expanded to $1.4 billion, and today, Cobre Panama is a cornerstone asset for FNV. FNV also adopted new innovations to streaming such as tying stream payments to a percentage of the gold price rather than a fixed price (Sabodala mine [Senegal] of Teranga

Gold Corporation in 2013) and doing syndicated streams (with Sandstorm Gold at the Karma mine [Burkina Faso] of Pure Gold in 2014).

Many investors want mining companies to financially engineer returns through the use of leverage and criticize lazy balance sheets. FNV's approach is to maintain a strong balance sheet and minimize the use of debt. In our view, there is enough cyclicality in the commodity sector without adding leverage. There is also an option value in having liquidity to make deals when others cannot.

The FNV approach was vindicated with the commodity downturn of 2014 to 2016. It forced even the largest global mining companies to focus on repairing their balance sheets. FNV was able to invest $1.8 billion in those years, creating foundational precious metals streams at Candelaria (Chile), Antamina (Peru), and Antapaccay (Peru). For the latter two transactions, FNV pioneered the first stream structures that allowed the operators to avoid debt treatment by the rating agencies. These large assets along with Cobre Panama have further diversified and strengthened the FNV portfolio with longer duration and higher quality assets than can be typically found in the gold industry.

Portfolio diversity is also a key focus for FNV. After the IPO, FNV was able to grow precious metals revenues enough to dilute its western Canadian oil revenues to under 20% of its revenues, which is an optimal weighting that FNV continues to maintain. By 2016, the overall portfolio had grown large enough to permit additional energy asset acquisitions and still stay within that targeted weighting. The opportunity was used to diversify the energy portfolio with gas and natural gas liquids (NGLs) rich royalties in Texas, Oklahoma, and Pennsylvania. Iron ore was also added to the revenue mix with effective royalties on Vale's iron ore systems in Brazil and Rio Tinto's Carol Lake iron ore mine in Labrador (Canada). FNV also owns a portfolio of royalties on copper development projects such as Alpala (Ecuador), Rosemont, Taca Taca (Argentina), and NuevaUnión (Chile). All these assets strengthen the diversity, duration, and quality of the overall portfolio.

While the big assets get the publicity, FNV has not forgotten its royalty creation roots. It is active both in buying early stage exploration royalties and in creating new royalties on promising geology. When FNV went public in 2007, it counted 190 mining assets. Today, it has 324 mining assets with most of the growth coming from these acorn investments. They do not get a lot of publicity as the initial investments are generally too small in relative valuation for FNV to justify a press release or feature in its investor materials. However, the longer-term potential can become material as highlighted with the Detour royalty creation example in the FN story.

Some recent FNV royalty creation examples, which look very promising, include the Ring of Fire (Ontario, Canada) where FNV helped Noront Resources consolidate the land position in return for royalties and east of Timmins (Ontario, Canada) where FNV helped Noble Resources acquire the mineral rights on which Canada Nickel Company is now outlining a large resource. It takes a long-term perspective to plant these acorns for FNV's future. Management sometimes claims it is working to make future CEOs of FNV look good.

LESSONS LEARNED

- Capital allocation needs to be the focus. After being part of an operating company, we learned about all the distractions that come from running a large organization with multiple mines. It is better to avoid the distractions and focus instead on growth.

- Diversity makes a portfolio stronger. Mining is a tough business and different commodities and assets will go through challenging periods. The larger the number of assets and the more diverse the portfolio, the better.

- Do our own due diligence. Looking back, it is the streams that we acquired through mergers and acquisitions, such as with Gold Wheaton, which have disproportionately needed to have its value written-down. It is important to do our own technical due diligence and structure transactions up front to mitigate risks. It is too late once the stream is established.

- Avoid net profit royalties or NPIs. Operators interpret NPI to mean "no profit intended" and use every accounting gimmick possible to minimize payout. Only the greatest mines, such as Goldstrike and Hemlo, can generate enough bottom-line profit to make the NPI valuable.

- Manage our egos. It is our portfolio, business model, and team approach (including our board) that generates success and not the brilliance of any individual.

- Maintain an ownership culture that is aligned with shareholders. Fundamental is managing overhead and paying progressive and sustainable dividends. Subway tokens are our "Franco limos."

POSTSCRIPT

Today FNV has grown its portfolio to more than 400 assets with over 100 generating revenues. FNV is the most valuable royalty company, the most liquid, and the most diversified by asset, operator, commodity, and geography. We are proud of our high margins, low overheads, and leading environmental, social, and governance (ESG) rankings. We take great satisfaction in the wealth that has been created for our investors since our $1.3 billion IPO in 2007. FNV has since paid out over $1.7 billion in dividends with per-share dividends increasing each of the past 15 years. Its market capitalization has reached more than $30 billion, ranking it among the most valuable gold companies in the world. This is against a book value of less than $6 billion, demonstrating real shareholder value creation. Since its IPO, FNV's share price has appreciated more than 11-fold and has generated an almost 20% compound annual growth rate return to its shareholders outperforming gold, its peers, and all the relevant market indexes.

Going forward, we continue to showcase FNV as "the gold investment that works" for our shareholders, our operating partners, and our communities. For our shareholders, we believe that combining a lower-risk gold portfolio with a strong balance sheet, progressive dividends, and exposure to exploration optionality is the right mix to hedge market instability. For our operators, FNV strives to be a good partner as it appreciates the challenges of mining. When the economics of an individual operation are challenged, FNV has cooperatively amended its royalties or streams. FNV has never needed a legal court case with any operator. FNV is proud to sponsor industry awards and support industry associations. For our communities, FNV starts by allocating capital only to responsible ESG projects. FNV is also working with our operators in growing its community programs at our portfolio assets. FNV has also played a leading role in helping the industry adopt a set of ESG standards now known as the World Gold Council's Responsible Gold Mining Principles. FNV has a vested interest in the success of the industry.

We are also being careful to preserve the DNA of our success, which stems from an ownership culture. In 2020, Brink, who was part of the FNV IPO and had led business development for 13 years, became CEO. Lassonde became chair emeritus, and I took on the role of chair. We all maintain large ownership positions in FNV. With FNV's new leadership, its diversified and high-margin royalty portfolio, and strong balance sheet, we couldn't be more optimistic about the future.

Finally, on a personal note, one of my CEO career highlights was the opportunity to meet the Duke of Edinburgh, Prince Philip, at Buckingham Palace. When he shook my hand and asked me what I did, I was proud to be able to say, "Your Royal Highness, I am in the royalty business too!"

This chapter is used with the permission of Pierre Lassonde and David Harquail.

3

ROYAL GOLD

NASDAQ: RGLD

royalgold.com

Stanley Dempsey Sr. and Tony Jensen

THE BEGINNING BY ROYAL GOLD'S FOUNDER AND FIRST CEO

Stanley Dempsey Sr.
This is the story of Royal Gold. It is a tale of loyal colleagues, technical excellence, luck, and occasional major corporate perfidy.

The early 1980s were tough for the mining industry. Commodity prices were in the tank. *Business Week* magazine declared "The Death of Mining" in a cover story.[1] The only bright spot was gold, which was hitting new highs after decades of being out of favor.

Many of us in mining decided that the future lay in precious metals. Ed Peiker, an engineer I had worked with at AMAX, and I decided in 1985 that we would organize a firm to seek opportunities in precious metals. We recognized that for the first time since the uranium boom of the 1950s, conditions were right for starting a new mining company. Those conditions included a rising price for a commodity, availability of projects, and investor interest. With robust gold prices, nearly every old gold mine, dump, and tailings pond around the world was prospective, and investors were eager to take a chance on gold.

---

1. "The Death of Mining," *Business Week*, December 17, 1984.

We organized a merchant bank, naming it Denver Mining Finance Company (DMFC). This gave us a way to earn working capital by way of fees for putting together mining deals. We had both been involved in all aspects of the industry and had enough financial background to be successful in dealmaking. One of our clients, Royal Resources, a NASDAQ-listed oil and gas exploration and production firm, used DMFC to shift to the gold-mining business. They sold their oil and gas assets, and DMFC put the proceeds into the Camp Bird mine in Colorado (United States) and the Colosseum mine in California (United States). They ended up buying DMFC from us and asking Peiker and me to run both the renamed Royal Gold and DMFC.

This bold switch from oil and gas to gold left us with a listed public company, a clean balance sheet, and a merchant bank that could earn both hourly consulting and success fees. We could use Royal Gold stock as part or all of the consideration for acquisitions.

Our initial strategy for Royal Gold was to buy working interests in start-up or operating properties. We also undertook exploration. In some cases, we joint ventured projects with major mining companies. We managed to parlay this into interests in four mines and six exploration projects.

Despite our quick start building up ounces of production, opportunities came along to also make quick gains by selling the mines we had just bought. We exercised a put at Colosseum and sold Hog Ranch (Nevada, United States) to Western Mining Corporation (WMC). Together, they gave us a gain of $6.1 million.

This performance confirmed that we were good dealmakers and explorers, and that we could operate mines. But it underestimated the costs of developing small mines and meeting environmental and safety regulations. It also failed to recognize that small mining companies are sometimes pushed around by bigger miners.

We supported our operating "habit" by making deals that kept us in cash. We were better at buying and flipping mines than we were at operating them.

Peiker and I had spent most of our careers with AMAX, one of the largest mining firms in the world. As an officer of AMAX, and as counsel for several of its divisions, I was directly involved in making sure that we obeyed the law and met all of our obligations to counterparties. Call me naïve, but I was shocked by the treatment that Royal Gold sometimes received from major companies.

I will avoid a who-hit-who portrayal of the events that occurred as we put Royal Gold on the road to success in the late 1980s and early 1990s. Eventually, Royal Gold obtained all of the relief that we sought when threatening litigation or actually filing suit. In all cases, we received satisfactory settlements. In recounting just enough of the story to make our history understandable, I do not wish embarrassment to anyone involved. I believe that actors in this drama have reconciled their personal grievances, and that we are friends once again. In this wonderful worldwide community we call the mining industry, friendship is the hallmark of our calling.

The last months of the decade of the 1980s exposed Royal Gold to a perfect storm of troubles. The price of gold weakened. The company announced on October 19,1987—the date of the Black Monday 1987 stock market crash—that samples from the Camp Bird mine had been tampered with, and that the company was having liquidity problems. Our share price took a tumble.

Adding insult to injury, Royal Gold's joint venture (JV) partner Placer Dome U.S. (PDUS) downplayed Royal's only good news, a discovery in the GAS-2 drill hole at the JV's Crescent Valley (Nevada, United States) prospect, as they tried to squeeze us out of a world-class discovery. I'll let WMC management tell you about Camp Bird.[2]

> In May 1986 Bruce Kay completed an evaluation of the Camp Bird Mine in the San Juan Mountains of south-west Colorado. The mine had produced four million ounces from high-grade veins. An option was negotiated with the owner, Federal Resources, who

2. Martin Summons, *Mandarins and Mavericks: Remembering Western Mining 1933–2005* (Richmond, VIC, Australia: Hardie Grant Books, 2017).

worked the mine on a small scale, with Royal Resources (to become Royal Gold) and Ouray Partners having 20% and 5% interests in a joint venture respectively, to develop additional reserves in the existing mine and explore for a new ore position on the Camp Bird vein to the east. ...

WMC were managers but retained all of Federal's existing mining and geological staff, including the Chief Geologist and the Mine Manager. ...

In 1986–87 and 1987–88, rehabilitation of the underground mine was in progress and underground exploratory drilling commenced. High-grade gold intersections were made. In 1987–88, 1,348 ounces of gold and 15,701 ounces of silver were produced from remnant mining. In 1988–89, the production was 2,022 ounces of gold.

Drilling of cross-veins, particularly the Monument, indicated high-grade gold in the undeveloped upper levels of the mine. Development reached the area in June 1989, but no gold was found in the vein. Re-assaying of all high-grade intersections of the Monument vein in August and September established that all high grades had been falsified. ...

WMC withdrew from the joint venture in October 1989, after settling a claim by Royal Gold for $3 million, and withdrew from management by late 1990 after posting a $100,000 environmental bond with the Colorado State Government. The net WMC expenditure, including the settlement and the bond, was $19,954,000.

The WMC involvement in Camp Bird was an unmitigated disaster. Instead of being an early source of cash flow, the project consumed a substantial sum. We were obviously extraordinarily naïve in trusting the previous management and allowing them to operate without any supervision. There must have been considerable amusement about the gullible Australians.

Roy Woodall, an officer and director of WMC tells his view of the Camp Bird story in a Bancroft Library interview.[3]

> By 1986, we had very good geologists working for us in the United States, and their mandate was to find us a gold mine. Our man in Denver found out that a once famous Colorado gold mine called the Camp Bird, high in the San Juan Mountains, was available. Yes! So, we had a look at the opportunity of reopening this mine, which is in one of the great mineral districts of Colorado. And here was a mine with a history of high-grade and a mine with an opportunity to find other lodes which might have never reached the surface. It looked perfect.

> …So we financed the refurbishment of the mill and reopened the mine. There was unmined gold ore known, and with the price of gold the way it was in 1986, which was at nearly $400 an ounce, it was economic. … So, we started the mill and we started producing gold and we started exploring.

> So here we were having made the type of discovery we were looking for; a new lode, and high-grade. We cross-cut to the lode, then we started to drift along it to really sample it. But we couldn't find any significant gold. So we went back to the drill core and did new assays and there was no gold in the drill core. So here is another tragedy of the eighties. … The samples which went for assay were "salted," i.e., deliberately contaminated with gold from another source. …

> Now, we had a junior company as our partner in this venture because we thought it was good to have a local Colorado company with us to keep us politically smart. Although these new

3. Roy Woodhall, "Roy Woodall: Australian Geologist, 1953 to 1995: Success in Exploration for Gold, Nickel, Copper, Uranium, and Petroleum," interview by Eleanor Swent, 2004, Oral History Center, The Bancroft Library, University of California, Berkeley, 2006.

drill intersections into a new lode were significant, they weren't so significant to Western Mining as they were to the little Colorado company. To them these were very significant intersections, and when they were announced, the price of their shares started to rise. …

The other company was Royal Gold. Now, I am quite sure that the directors of Royal Gold were as disappointed as we were. But they had a problem also because they had announced false assay data to their shareholders. So, we not only wasted good money following up false assay results, but we had to pay compensation to Royal Gold for alleged bad management. I guess from their shareholder's point of view that was fair enough. Let's leave the 1980s. So many "set backs," so many lost opportunities, some perhaps best described as disasters.

The saga of our decade-long struggle to hang onto our interests in the Cortez mine (Nevada, United States) is long and involved, but here is the short version.

In 1987, ECM of Billings, Montana (United States), began staking claims (which they called the GAS claims, for Gold Acres South) on the pediment just outside the Gold Acres mine. Some of this ground was included within lode and association placer claims held by Cortez Gold Mines (Cortez or the Cortez JV, majority owned by PDUS). As the ECM contract stakers moved northward, they ran into crews from Gold Fields Limited, who were staking from north to south. Gold Fields ended up with a claim block just east of the Gold Acres heap leach facility, and ECM secured possession of several thousand acres of ground immediately southeast of the Gold Acres mine.

ECM recognized the potential for sediment-hosted gold deposits in the pediment east of Gold Acres and moved quickly when they found that the land was open for entry. Subsequent to staking the ground, ECM approached Royal Gold to see if we would be interested in farming in. We liked the address in the Crescent Valley. ECM's approach was also timely, as we were still in a position to raise money for exploration.

We had two reservations. First, we did not want to be involved in a claim conflict. Second, we were worried about the depth to bedrock. Our first objection went away when PDUS indicated it might be willing to work out a compromise on the claims if someone like Royal was willing to spend some serious money on exploration. We resolved the second problem by doing geophysical work to get a handle on depth to bedrock.

Royal Gold leased the property from ECM and then turned half the deal to an Australian firm with which we had good relations, Golden Bounty Resources. Royal next entered into an agreement with Cortez, whereby Royal would do the first million dollars' worth of exploration on the GAS claims. If the JV was successful, Cortez would put the mine into production, and Royal and Golden Bounty would have a 20% carried interest. Cortez would build and operate the mine. The agreement also provided a mechanism for settling the claim dispute. Subsequently, Royal Gold, as operator on the GAS claims, performed several phases of exploration, finding detectable gold in nearly every hole, with encouragement in the second hole, GAS-2. This and subsequent work by Royal was done under the direction of Roger Steininger. Joe Anzman did much of the geophysical work. On February 1, 1988, we announced that "gold mineralization has been encountered within a trace element anomaly that may be indicative of a sediment-hosted gold deposit."

Our next drilling campaign consisted of 15 reverse-circulation holes, totaling more than 8,400 feet in September 1988. On October 28, 1988, Royal announced that "additional and significant gold mineralization has been encountered. ...The mineralization is a continuation and expansion of that encountered in drilling last year."

Our last drilling effort was undertaken in the spring of 1990, and it produced some very attractive results from additional drilling on a grid around GAS-2.

In the years post–Black Monday, the price of gold and of our shares came down. We were having trouble funding our exploration projects. This problem was made worse when Golden Bounty advised they were also having trouble funding their half of the JV. We eventually (May 4, 1990) purchased Golden Bounty's interest for shares of Royal Gold.

We struggled mightily to stay current and to do the work required of us by the ECM lease. ECM worked with us on various restructuring arrangements, and together we were able to keep the project moving forward.

We tried for a year or so to restructure our deal with Cortez. They were not cooperative. In 1991, we ended up selling our interest in the project to Cortez.

In late 1991, Cortez announced the Pipeline discovery in Nevada (United States). Pipeline is about one-half mile north of the GAS claims. By May 1992, Royal Gold, Gold Fields, and ECM all ended up in various lawsuits with Cortez over the events surrounding the sale of interests in claims that encompass or are nearby to the Pipeline deposit. Happily, Royal Gold quickly reached a settlement with Cortez. The settlement agreement restored our 20% net profits royalty, and provided that

> If Cortez does not elect to put the Project into production by early 1996, then Royal may elect to put the property into production by granting a production royalty to Cortez identical to the one described above. Royal would then be entitled to use, under a normal tolling arrangement and as available, the Cortez milling facilities in the vicinity, including any to be built for the Pipeline discovery.

This so-called flip-flop arrangement proved troublesome to PDUS, and renewed negotiations in 1999 resulted in a conversion of our net profits interest in South Pipeline into gross smelter return royalties extending over the mining complex that includes the Pipeline and South Pipeline gold mines in Lander County, Nevada (United States).

Given the situation that Royal Gold was in in 1991, we took a hard look at alternative futures, including exploration and operation, and decided to adopt the royalty firm model. I often tell people that we decided to become a shameless copy of Franco-Nevada Corporation and Euro-Nevada Mining Corporation, two highly successful Canadian royalty companies. And in many ways, we did just that. Franco-Nevada led the way, showing that investors had an appetite for royalty income and that they would pay more for that kind of income than they would for income from operating gold-mining firms.

Royal Gold sold or shut down its operations and put its entire effort into acquiring and managing royalties and developing exposure to exploration opportunities. We were fortunate to have a marquis royalty at the Cortez mine in Nevada (now a JV between Barrick Gold Corporation and Newmont Corporation), which started furnishing significant revenue in 1992. Since then, we have added many additional cash-generating royalties, by acquisition or by financing mine development in return for a royalty.

Having made the conversion to the royalty model, and now enjoying revenue from Cortez and a higher share price, one of our early acquisitions was High Desert Mineral Resources (HD). This was an exploration firm owned by recently widowed Lee Halavais. Halavais and her husband discovered the Leeville mine on the Carlin trend in Nevada (United States). This intrepid team aspired to creating a royalty firm and had sold the mine to Barrick, reserving a royalty on Leeville, and also receiving a royalty on the Goldstrike mine (Nevada, United States).

We were on good terms with Halavais, having been in a JV with HD. We began negotiations with her to acquire HD. Franco-Nevada stepped up with a bid, but our offer of our shares was accepted. We were happy to secure these high-value royalties. We were no longer a one-trick pony!

With ever greater royalty revenues coming in, the story in the early 2000s was one of acquiring new royalties, securing an ever-larger line of credit and a series of equity offerings. Royal Gold paid its first dividend in 2000.

Our successors at Royal Gold have taken our work and improved on it, building one of the largest gold firms in the world. Tony Jensen has attracted a first-rate board of directors and an astute management team. They have built the company far beyond my expectations, acquiring world-class royalties and streams and managing them effectively. They continue to provide leadership within the industry, supporting organizations like the National Mining Association and the World Gold Council. Their financial performance is remarkable, and I am happy that they have continued to provide a continually growing dividend. My 60-year-long career has brought me many opportunities and the pleasure of working with many of the finest people on the planet. The Royal Gold people are at the top of that list.

GROWTH BY ROYAL GOLD'S SECOND CEO

Tony Jensen

I was introduced to the royalty business while working with PDUS and Kennecott Minerals Company as the mine general manager at the Cortez JV in Nevada from 1999 to 2003. A separate book could be written about the royalty intricacies of Cortez, and the ole adage of "no good deposit is without a lawsuit" held true there.

At the same time of learning the royalty business from the payor perspective, I met Stanley Dempsey (Royal Gold's first CEO and founder) and Craig Haase (Franco-Nevada's then director, executive vice president, and chief legal officer). We quickly became close professional colleagues on all matter of issues while sitting on the board of the Nevada Mining Association. I admired the simplicity and efficiency of the royalty business after experiencing nearly 20 years of operating complexities.

Just after the discovery of the multimillion-ounce Cortez Hills deposit in 2003,[4] I stepped out of well-respected Placer Dome (then with a market capitalization of approximately $6 billion, sales of $1.8 billion, and about 10,000 employees) after 18 years to join Royal Gold, which then had a market capitalization of approximately $440 million, revenue of $16 million, and fewer than 20 employees. I never dreamed that Royal Gold would someday exceed Placer Dome's market capitalization (even in nominal terms) to become one

---

4. Placer Dome's 2003 annual report stated, "The Cortez joint venture discovered a major new mineralized zone called the Cortez Hills deposit. Placer Dome's 60% share of the estimated proven and probable mineral reserve is 3.2 million ounces of gold (22.5 million tonnes with an average grade of 4.36 grams per tonne)... Exploration work at the deposit, which is open along strike and at depth, is ongoing." (Placer Dome, "2003 FY Annual Report, Canada," U.S. SEC Form 40-F, 66.)

of the top 15 largest precious metals companies in the world by my retirement at the end of 2019.

Interestingly, Franco-Nevada had given up on the royalty sector as a pure play in early 2002 when they sold the business to Newmont, and Repadre Capital Corporation did the same via a transaction with IAMGOLD Corporation in late 2002. While leaving the security of Placer Dome was risky, the timing for an entrepreneurial career move could not have been better. The gold market was at the start of what turned out to be a 10-year bull run, and Royal Gold had the sector largely to itself.

Learning One: Premium trading multiples for the royalty sector afford a lower cost of capital and discount rates.

One of the early learnings in the royalty business was the significantly lower discount rate used to value acquisitions with respect to those used for operating company investment hurdles. We analyzed our cost of capital with investment banks, and some even argued our cost of capital was negative because of our equity trading multiples. The market generally rewarded Royal Gold in the 20 to 30 times cash flow or two times net asset value range, a significant premium to most gold producers.

The reason for this premium is clear; Royal Gold had the characteristics of a large cap company in a small cap entity with plenty of room to grow. Specifically, the royalty business model has much of the upside of large producing companies (diversity in revenue source and geography, high-quality assets, exploration and price appreciation option value, production increases potential, strong balance sheets) but at a much lower risks (no production cost increases, no capital cost overruns, no management complexities, no environmental liabilities, and very low overhead).

This lower cost of capital is key to success in the royalty and streaming business. We could buy assets from producing companies at a price far superior to their internal valuation. When we ventured into copper deposits to acquire precious metal by-products in the 2010s, the value arbitrage was even greater with base metal companies.

We had good success in the 2000s building our company. We focused primarily on purchasing existing assets, and the opportunities were plentiful. We purchased royalty portfolios from Kennecott Minerals, Rio Tinto Group, AngloGold, and Barrick Gold. At the time, major mining companies did not consolidate or value their royalty portfolios, so it was only natural to sell these non-core assets but getting their attention among other priorities was a challenge. Since then, every mining company has kept a keen eye on their internal royalty portfolios and have actively marketed them from time to time as a source of capital. We also acquired interest in royalties from private parties, and there was little competition in the early 2000s.

Unfortunately, success does not go unnoticed, and it always invites competition in a free market. And in the royalty business, the barrier to entry is low if capital is available to experienced mineral dealmakers. Battle Mountain Gold Company and International Royalty Corporation were active in the sector. Silver Wheaton Corp. was spun out of Goldcorp in late 2006 but was initially focused on only silver transactions, and Franco-Nevada was reintroduced into the public market in late 2007 with renewed vigor to focus solely on the royalty sector.

We conducted annual extensive strategic sessions and always challenged the business model and its relevance in the future. We questioned the growth potential and generally felt that the business would struggle to grow once we obtained $100 million in annual revenue. The concern was for the lack of sizable existing royalties that could potentially come to market. We realized then that we had to create our own royalties and chose the shorter route of project finance rather than the long duration and risky route of exploration finance. We did our first royalty financing at the Troy mine in Montana (United States) in 2004 and followed that up with royalty financing for the Taparko mine in Burkina Faso in 2006. In late 2004, Silver Wheaton (at the time, now Wheaton Precious Metals Corp.) began the innovative stream-financing model. As that product matured, we considered the attributes it had for U.S.

tax regulations and shamelessly adopted the model in 2010 when we provided financing for the Thompson Creek Metals Company acquisition of Terrain Metals Corporation and subsequent project financing for the Mount Milligan mine in British Columbia (Canada). Royal Gold has only completed stream financings since, and growth concerns were addressed with the ability to create new products.

We were active in consolidating the industry in the second half of the 2000 decade. We negotiated a non-contested transaction with Battle Mountain Gold in 2006, and in 2010, we were able to close a friendly transaction with International Royalty after notable competition from Franco-Nevada. Battle Mountain and International Royalty suffered from limited balance sheets in an environment where deals were getting larger, which brings me to a second early learning of the royalty business: A strong source of capital or cash flow is required to fuel the business. Our early fuel was Cortez, Franco-Nevada's was Goldstrike, Wheaton Precious Metals was incubated within Wheaton River Minerals (and later Goldcorp) to purchase the Luisman (Mexico) and Zinkgruvan streams (Sweden) prior to their spinout, and Triple Flag Precious Metals Corp. had the backing of Elliot Investment Management. Those sources of fuel certainly accelerated the growth trajectory of these companies versus other entrants that took much more time to gain critical mass.

Learning Two: *A strong source of capital or cash flow is required to fuel the business.*

Although Royal Gold has been and continues to be very successful, it is difficult not to think about the ones that narrowly got away. In the fall of 2007, we spent endless hours locked away in our lawyer's conference room negotiating with Newmont for the sale of the Franco-Nevada royalty portfolio. Newmont decided to change directions under new management and initiated a dual-track process for either the potential sale or a public spin-off of the royalty portfolio. Newmont knew that the buyer universe at the time for a sale was limited, thus the dual-track process. Our offer of substantial cash and Royal Gold shares was competitive, but a strategic sweetener was the contribution of the Leeville royalty on Newmont's mine in Nevada. It was a bit of a stretch for Royal Gold to buy a set of assets priced at about 25% more than our market capitalization, but the reward was worth the risk and we pursued the acquisition and financing aggressively. While the public spin-off was being pre-marketed to potential investors in appropriate jurisdictions, we completed final agreements with the Newmont sales team. On November 19, 2007, Newmont President and CEO Richard O'Brien had just returned from a flight from Europe and asked that Dempsey and I meet him in his office in the late evening. His principal message was that if we closed by year end, we would sign the purchase and sale documents that evening and announce the transaction in the morning. That sounds simple enough, but we were incredibly disappointed that we could not convince our bankers to acceler-ate the financing essentially one month. The world later learned what banks had likely already started to become aware of, which is that we were already entering the Great Recession, undermining the solvency of several in the financial sector. Which brings me to my third learning: Always hire invest-ment banks that have strong balance sheets and are willing to put their own capital to work. We offered to modify the total consideration to increase the share component thereby compensating for any Newmont year-end 2007 tax issues, but that did not prove successful. The equity spin-off was announced shortly thereafter.

Learning Three: *Only hire investment banks that have strong balance sheets and are willing to put their own capital to work.*

I should reference another miss that resulted in an evolution in the busi-ness. In early 2012, we had been in extensive discussions with Inmet Mining Corporation to finance the construction of the Cobre Panama project in Panama. This was a highly competitive piece of business that ultimately went to Franco-Nevada, consisting of a $1 billion stream—the largest stream to that date. Royal Gold was well positioned to win the business, but we had a financing condition, which was common in the sector up to that point in time. This brings me to a fourth learning: Always be in the position to buy. We did not have financing conditions in our proposals thereafter.

Learning Four: *Always be in a position to buy.*

At the end of the decade, the metals supercycle was racing, and gold peaked at nearly $1,900 per ounce in 2011. Major mining companies invested heavily and faced huge project capital overruns during the supercycle. By the middle of the decade, copper and gold prices dipped to $2.00 per pound and $1,050 per ounce, respectively, and aggressive refinancing was required throughout the industry including Vale, Glencore, Teck Resources Limited, and Barrick. By that time, stream financing was well understood by the mining sector. Royal Gold entered new deals totaling $1.4 billion from May to August 2015. This was an extraordinary time for the business, and our competitors were similarly active. We were in a strong position to buy when all this business broke loose.

I feel compelled to write about the most significant challenge Royal Gold faced as a company during my tenure as president and CEO. In 2015, the owner of our largest revenue source, the Mount Milligan mine, was experiencing financial difficulties and the market grew very concerned about Royal Gold's ability to maintain its business interest. After the flurry of investments just described, the company focus turned 100% on protecting our stream at Mount Milligan.

We retained highly capable investment bankers and bankruptcy experts in the United States and Canada to fully understand the playing field and to know our options when Thompson Creek's restructuring advisers came calling. We developed a plan to market Thompson Creek Metals and/or the Mount Milligan mine to the precious and base metal industries. The work was called

Project Chess because of its strategic value to Royal Gold and the strategic moves that would be required to ensure that the stream was protected. We developed pitch books and visited with scores of companies about our interests and the modifications we were willing to consider. Several companies showed interest by the end of the year, but one—Centerra Gold—went from a low knowledge base to winning an agreement with Royal Gold over one weekend.

In February 2016, Royal Gold agreed in principle with Centerra to reduce its gold-only stream in return for a new copper stream on Mount Milligan to make the total stream interest more attractive to a precious metal producer. Soon thereafter, Thompson Creek Metals began a formal sale process, and our previously interested parties were subsequently tied up with confidentiality agreements. As such, all avenues of communication were properly terminated.

In late June 2016, Jacques Perron, the president and CEO of Thompson Creek Metals at the time, called to say that Thompson Creek had come to terms to sell the company to Centerra and that he understood that we had an agreement in principle with Centerra regarding the Mount Milligan gold stream. He informed Royal Gold that they planned to close the transaction in two weeks partially conditional on Centerra and Royal Gold entering a definitive agreement. Because we had largely completed our negotiations months earlier, the pieces quickly fell into place, and the deal was announced in early July. This was a complex and intense strategy that was rewarded. Royal Gold's share price went from $25.33 on January 1, 2016, to $83.22 on July 11, 2016. This result was because of the successful completion of a well-planned strategy by extremely competent staff and advisers, which brings me to my fifth learning: Always hire the best in the business.

Learning Five: *Always hire the best in the business.*

The royalty and stream sector has long differentiated itself from the rest of the precious metals market. Characteristics, such as strong balance sheets, robust free cash flow, consistent and longstanding dividend policies, low loss ratios on investments, efficient use of capital, low operating expenses, and attractive total shareholder returns, were not attributes commonly thought of for the precious metals business. I enjoyed speaking with new potential investors and always emphasized that Royal Gold was an investment for all investment styles, not just for gold bugs.

The secret sauce is the premium afforded to the royalty and stream business model. Without that, there would be no attractive cost of capital and no value arbitrage in acquiring new assets. Our acquisitions were targeted at protecting and enhancing the premium. As such, we resisted change to the business model and did not give up optionality in our deal structures. Although evolution of the business model was inevitable, changes to deal structure were carefully considered with the understanding of potentially setting new precedence in future deals.

My entire career was blessed with some incredible experiences with excellent people. Seeds of friendship were scattered throughout the world, and those seeds have taken root everywhere. The people of the mining industry are extraordinary in many ways, and the leaders of the royalty and stream sector are much more than just fierce competitors. They are respected colleagues, as represented by cooperation of producing this book, and many are lifelong friends.

4

WHEATON PRECIOUS METALS

TSX | NYSE | LSE: WPM

wheatonpm.com

Randy Smallwood

WHEATON AND THE HISTORY OF STREAMING

In the early 2000s, a group of executives forever changed the world of mine project financing with the development of an alternative form of funding that would become known as *streaming*. This started the journey to create what is now one of the world's largest precious metals companies by market value, changing mine financing forever.

Those who aren't familiar with the history of Wheaton Precious Metals Corp. may be surprised to learn that the Wheaton name can be traced much further back in time to the historic gold-mining district of Mount Skukum in Canada's Yukon Territory. The original parent company, Wheaton River Minerals (Wheaton River) was incorporated in 1990 to explore and develop Mount Skukum, located in the Wheaton River valley, Yukon Territory (Canada). The story of its predecessor is studded with decades of innovation and strong growth and makes for an interesting prelude.

The Wheaton reputation for innovation began in a small corner of the Golden Triangle, in northwestern British Columbia. With a lack of success exploring the Mount Skukum project, Wheaton River acquired the past-producing Golden Bear mine (British Columbia, Canada) from Homestake Mining Company in 1993 and immediately embarked on an aggressive exploration campaign. Success was quick, and by 1996, the mine was ready to restart, but this time with a unique operating system.

Golden Bear Mine: Seasonal Heap Leach

Wheaton River Minerals operated the Golden Bear mine as a seasonal heap-leach mine, at an elevation of 1,800 meters in northwestern British Columbia from 1996 to 2001. With winter snow depths exceeding 9 meters, avalanche risk forced the mine to operate seasonally, with snow clearing beginning in March and stacking operations in April. After a very intense summer campaign, stacking would stop in early October, and solution processing would wind down through November. Golden Bear was very profitable, producing up to 100,000 ounces of gold per year at cash costs as low as $100 per ounce.

The Golden Bear mine might well have been called the Grubstake mine, as the profits from Golden Bear built up a very attractive, undervalued balance sheet for Wheaton River.

The modern Wheaton River journey began in 2001 when several mining entrepreneurs—Frank Giustra, Neil Woodyer, and Eugene McBurney—decided to transform what was a junior mining company with an incredibly undervalued balance sheet into "the best gold company in the world."[1] To achieve this ambition, they took over the Wheaton River board and brought in Ian Telfer as president and CEO to lead the company.

I originally joined Wheaton River as an exploration geologist for the Golden Bear mine in 1993, became the project manager within the successful Golden Bear team, and am now the president and CEO of Wheaton. When Telfer

---

1. Samantha Reynolds, *Out of Nowhere: The Wheaton River Story* (Vancouver: Echo Memoirs, 2008).

joined Wheaton River in 2001, he invited me to move to the head office as vice president of corporate development and lead all project evaluations.

In late 2001, Wheaton River began what was to become a mammoth expansion and consolidation campaign, with the first transaction as the $75 million acquisition of Luismin, the non-core mining division of Sanluis, itself a Mexican-based auto parts manufacturer.

The First Step in Growing Wheaton River

On its acquisition by Wheaton River Minerals in May 2002, the Luismin division of Sanluis ranked as Mexico's third-largest silver producer and the second-largest gold producer. Luismin comprised three mining operations in Mexico: San Dimas, San Martin, and La Guitarra. San Dimas itself consists of three mine areas centered between Durango and Mazatlan in the Sierra Madre range. In 2001, the three operations collectively produced 98,000 ounces of gold and 5.8 million ounces of silver, which is 190,600 gold equivalent ounces, at a production cost of $200 per gold equivalent ounce.

By 2004, Wheaton River, through a series of additional acquisitions, had rapidly grown into the fourth-largest gold producer in Canada and was aggressively looking to play a part in further industry consolidation. Although the company was focused on gold, its mines also produced significant by-product metals, including copper and silver. In fact, in contrast to Wheaton River's original mission, more than 40% of its revenue came from these by-product metals, and investors were beginning to question whether the focus on gold was genuine, and whether Wheaton River should be treated and valued as a gold company. One of my earliest mandates in this position was to purify the revenue streams of the multi-commodity-producing mines in the company's expanding portfolio.

The problem at Wheaton River was that, while trying to build "the biggest and best gold company in the world," we had invested in projects that were multi-commodity focused. So, by the end of 2003, about a third of our revenue was derived from the copper at Alumbrera (Argentina) with another significant percentage being attributed to silver, mainly from our Mexican operations.

The inspiration for what the mining world knows today as the streaming financing model was born out of a proposed copper/gold swap deal at the Alumbrera mine, which was jointly owned by Wheaton River and Mount Isa Mines Limited.

Telfer kicked this off and we sat down with Mount Isa Mines. We spent a fair amount of time negotiating a potential deal with them, whereby we would trade or "stream" our attributable portion of copper from the Alumbrera mine to them, and they would trade or "stream" their attribution portion of gold to us. It was through these negotiations that the concept of streaming came to life. However, despite months of negotiations, the parties eventually walked away without a deal.

Early in 2004, while the Wheaton River management team was sitting around the boardroom table, the streaming idea was raised again, but this time in the silver space. At that time, none of the listed silver producers were very profitable, given that the silver price had traded in a narrow range between $4 and $6 per ounce for several years. Despite this lack of profitability, these silver mining companies were trading at very high net asset value multiples.

I remember Telfer commenting on the valuations, saying, "imagine what multiple a profitable silver company could trade at."

Wheaton River's Luismin subsidiary was producing large amounts of silver, particularly from San Dimas, along with gold. In 2003, Luismin produced 106,900 ounces of gold and 6.09 million ounces of silver.

Given the large quantities of silver, Wheaton River thought it could extract optimum value from the silver being produced within its company by selling that silver to a subsidiary set up internally, which would increase the visibility of that silver production. Furthermore, it believed that by trading independently as a pure silver producer, the subsidiary could achieve that premium valuation that silver miners were receiving, or perhaps even better, given that it would be profitable.

Thus, the streaming model took flight through the subsidiary company. Silver Wheaton Corp. was incorporated in October 2004, and the streaming model was created, streaming all the silver production from Wheaton River's Luismin mines in Mexico. On its launch, Silver Wheaton was the only pure silver mining company in the world, as all other silver companies had an associated metal production such as gold or lead. Thus, a unique and exciting investment proposition was created, in which Wheaton River retained about a 75% stake.[2]

Wheaton's Streaming Model

The streaming model, initiated by Silver Wheaton (now called *Wheaton Precious Metals*) and now utilized by many other companies, allows a mining company to crystallize the value of future precious metals production from an operating mine or near-development project. The precious metals production is often the by-product of a base metals asset, but can also be a core product, such as gold from a gold mine. In return for an up-front payment made to the mining company, the streaming company gets the right to purchase a percentage of the mine's precious metal production for a predetermined amount on a per-ounce basis, typically for the life of the mine.

Streaming agreements generally do not include an ownership stake in the mine, nor do they involve the streamer having any operational control. As such, Wheaton focuses on streaming deals

2. Wheaton Precious Metals, "Silver Wheaton Transaction Completed," news release, October 15, 2004, www.wheatonpm.com.

on mines or projects that are low cost, high margin, and sustainable, meaning they have a good social license to operate.[3]

By entering into a streaming agreement, traditional mining companies can receive greater value for their by-product precious metals than what is reflected in the market. These companies can use the up-front payment to continue growing their core business, either through exploration, production expansions, or acquisitions; alternatively, the cash received can be used to strengthen their balance sheet in a non-dilutive form.

For the streamer, the limited risk related to mining operations also provides considerable stability and upside potential. In Wheaton's case, the cost predictability embedded in the streaming model provides direct leverage to potential increases in precious metals prices. With ongoing operating costs set at the time a stream is entered into at a predetermined delivery payment, Wheaton can deliver among the highest cash-operating margins in the mining industry. Wheaton is also not responsible for any exploration costs or environmental liabilities, and it is not exposed to currency risk, as it receives metal from the producer, and sells that metal in U.S. dollars. While not responsible for environmental liabilities, as Wheaton grew over the years and the concept of sustainable development became far more entrenched in business practices, Wheaton has increasingly expanded its skills and expertise on mine environmental and social stewardship and best practice, lending this expertise to partners as and when necessitated.

In the first transaction of its kind, Silver Wheaton's subsidiary agreed to pay $36 million in cash and gave Wheaton River a 75% stake in the company

---

3. Gwen Preston, "The Silver Wheaton Success Story," *The Northern Miner*, February 1, 2010, www.northernminer.com.

for a right to buy all the silver to be produced out of Luismin, for $3.90 per ounce for a term of 25 years.[4]

When we went to paper that first streaming deal, Telfer decided that $3.90 per ounce was more appealing than $4 per ounce, which was then San Dimas's hard cost. With a 1.5% inflation kicker that started in the fourth year, this became our production payment standard.

To fund the deal, which was valued at $209 million, Silver Wheaton issued 175 million new shares (along with 87.5 million warrants at $0.80) at a price of $0.32 per share to raise $56 million. Using that base price, Wheaton had an initial market value of $232 million.[5]

Strong Growth

Posting a 5-for-1 stock consolidation in December 2004, the implied starting share price was $1.19 per share. To put that into perspective, Wheaton's share price closed at $42.93 at the end of 2021 and it paid total dividends of $0.57 per share over the year.[6]

The timing of the transaction could not have been more ideal as silver prices would see sizable increases over the next three and a half years from a base of $7 per ounce in October 2004, spiking at more than $45 per ounce a few years later.[7] Production of silver at the Luismin mines also just kept increasing.

---

4. "Wheaton River Minerals Ltd. and Chap Mercantile Inc.: Silver," *Bloomberg* US Edition, October 15, 2004, www.bloomberg.com.

5. Wheaton Precious Metals, "Chap Mercantile 175-Million-Share Private Placement," press release, October 20, 2004, www.wheatonpm.com.

6. Wheaton Precious Metals, "Silver Wheaton Shares to Commence Trading on a Consolidated Basis," press release, December 20, 2004, www.wheatonpm.com.

7. "Silver Prices—100 Year Historical Chart," Macrotrends, updated September 2022, www.macrotrends.net.

Meanwhile, in December 2004, shortly after Silver Wheaton was created and the silver stream transaction was concluded, the Canadian gold major Goldcorp and Wheaton River announced a transaction to combine the two companies. The deal was successfully completed in early 2005, with management of the smaller Wheaton River taking over the combined entity, and creating a gold-mining powerhouse, Goldcorp.

At the same time, Silver Wheaton was in the process of concluding its second transaction, the acquisition of 100% of the payable silver produced by Lundin Mining Corporation's Zinkgruvan zinc operation in Sweden, where production was expected to average approximately 2 million ounces of silver per annum for over 20 years.[8]

Notably, as Zinkgruvan had been operating since 1857, entering into such a precious metals agreement brought a novel financing method to one of the world's longest running mines. This highlighted the opportunity that lay ahead for Silver Wheaton, not only to fund existing brownfield assets but also, as it was later proven, as a funding source for new mines and acquisitions.

Lundin Mining had just bought the Zinkgruvan mine from Rio Tinto Group for $100 million plus add-ons.[9] For the silver stream, $50 million in up-front cash and $25 million worth of Silver Wheaton shares were advanced to Lundin Mining. Ultimately, the shares tripled in value over the next six months, delivering nearly $125 million in total value back to Lundin.

Essentially, within a year, Lundin had sold off its non-core silver by-product production, and essentially acquired Zinkgruvan for free. Lukas Lundin, chairman of the company, recognized the opportunity this created and used that original silver stream to finance the growth of Lundin Mining. That really resonated through the marketplace, and streaming took off from there.

---

8. Wheaton Precious Metals, "Silver Wheaten Completes Zinkgruvan Silver Transaction," press release, December 8, 2004, www.wheatonpm.com.

9. Lundin Mining, "Lundin Mining Corporation: Second Quarter Report," news release, August 13, 2004, https://lundinmining.com.

Next up was a silver streaming deal with Glencore at their Yauliyacu mine in Peru, and it just kept on growing.

Goldcorp Exits

Over the years, Goldcorp had been slowly selling some of their shares in Silver Wheaton Corp., as the market was always worried about the overhang. In 2008, Peter Barnes, the CEO of Silver Wheaton from 2006 to 2011, reached an agreement with Goldcorp to sell its remaining stake in Silver Wheaton for approximately $1.47 billion, a significant increase in value considering the original $209 million deal value, a little over three years earlier.[10]

At that time, it was the largest secondary offering in Canadian history. In all, by the time Goldcorp sold the San Dimas mine to Primero Mining Corporation in 2009, more than $3 billion in value had been created from the Luismin Mexican assets, which were bought by Wheaton River Minerals in 2001 for $75 million.

As streaming was increasingly being recognized as an extremely attractive funding option, the deals got bigger and bigger. With the value and size of deals growing ever larger, Silver Wheaton was able to take advantage of these opportunities.

I officially moved from Goldcorp to Silver Wheaton as executive vice president of corporate development in 2007, after Telfer moved up from CEO to the chairman role at Goldcorp following that company's acquisition of Glamis Gold.

Within two years of my official move, two major opportunities materialized: the first in 2009 through the acquisition of Silverstone Resources, a smaller

---

10. Wheaton Precious Metals, "Goldcorp Completes Sale of Its 48% Interest in Silver Wheaton," press release, February 14, 2008, www.wheatonpm.com.

silver streaming company with three active streams; and the second, the signing of a major deal with Barrick Gold Corporation in the same year. The Barrick deal saw Silver Wheaton provide a $625 million up-front payment over three years, in exchange for 25% of the future life-of-mine payable silver production from Pascua-Lama (Chile/Argentina) as well as immediate production from Barrick's three South American operating mines.

Both deals had the effect of transforming Silver Wheaton into the then largest streaming company in the world with a market capitalization of over $6 billion at the time.

A Tight Ship

These days, Wheaton Precious Metals Corp. runs as a tight ship, with just 45 employees. As such, the company has consistently had one of the highest market capitalizations per employee on the New York Stock Exchange.

The company also prides itself in recognizing the deals it walks away from as much as the ones it closes. It takes a lot of strength to get in front of the deal freight train as it is coming down the tracks; it is hard to say no to a deal once it has gained significant momentum. To achieve this, we have to make sure that everyone in the company can freely voice their thoughts, beliefs, or reservations about a transaction at any stage of the evaluation process.

In 2010, I was appointed president and a year later, added the CEO title, taking over from the retiring Peter Barnes. It was at that point that we started asking ourselves, where we were going to take the company next. There were not as many opportunities in silver as there were in the gold space.

Up until 2012, Wheaton had signed only two transactions with agreed deliveries of gold: Hudbay Mineral's 777 mine (Manitoba, Canada) and

Constancia project (Peru), with the bulk of the value of those deals being in silver. This changed in 2013, when Silver Wheaton took a transformational step forward, completing three transactions, all of which were for gold.

The first of these, and still the largest ever streaming deal ever transacted, was with Vale on the Salobo (Brazil) and Canadian Sudbury (Ontario) mines at a value of over $1.9 billion. In addition to the advance cash, Silver Wheaton also issued warrants to Vale, which boosted the value of the deal. In total, Wheaton paid $3 billion in connection with the Salobo stream as Wheaton's interest, as well as the mine, continued to grow (Salobo is Brazil's largest copper mine).

Not only was it our biggest, but it was our first serious entry into the gold space. It was also an interesting one, because it really highlighted a key strength of Wheaton. The comments from Vale were that we won the bid, which was a competitive process involving all of the large streaming and royalty companies, because those competing against us only sent one or two internal people on the due diligence visits, along with a delegation of consultants. In our group, every single person on the due diligence team were employees of Wheaton, and actually had ownership in the decision. Vale took notice and appreciated the internal expertise Wheaton had developed.

And there was another important aspect of the Vale transaction. After the team had finished its due diligence review of Salobo, we believed Vale had been too conservative in several different areas of the resource and reserve estimation and thought the mined gold grade would be 10% to 12% higher than Vale was reporting.

How we value is based on what we believe. That is the underlying decision no matter what we do here—we will make own our decisions and will never point the finger at someone else. With mined precious metal grades often being higher than forecast, Wheaton's estimates on production have proven to be accurate in many cases throughout its history.

Each year, we look at every asset and come up with what we believe will happen there. On an individual asset basis, in the spirit of a good partnership, we will always align with our partners and their production forecast, but on our aggregated production number, that is our number that we need to make sure we get right.

By the end of 2016, Silver Wheaton had mobilized over $8 billion in funds and held streams in 22 operating mines and 8 development projects globally.[11] As the company's revenue was almost evenly split between silver and gold production, the Silver Wheaton name no longer represented its diverse portfolio of gold and silver assets. As a result, on May 10, 2017, Silver Wheaton officially changed its name to Wheaton Precious Metals. The new name reflected the diverse portfolio and reinforced its position as the leader in precious metals streaming.

Stock Exchange Listings

Importantly, the transparent nature of streaming as a business model, but also a funding one, allows for the company to be easily listed and investible on global stock markets. In addition, the unique nature of the product, with its low costs and long-life assets, makes it an enticing investment proposition.

It was no surprise then, that having graduated to the Toronto Stock Exchange (TSX) from the TSX Venture Exchange in 2004, the company also joined the New York Stock Exchange two years later.

In 2020, the company added a third listing on the London Stock Exchange to the fold, making Wheaton a truly global precious metals company.

---

11. Silver Wheaton, *2016 Annual Report* (Vancouver: Silver Wheaton Corp., 2016).

A year after changing its name, Wheaton further diversified its streaming portfolio, with transactions in cobalt and platinum group metals.

In June 2018, Wheaton concluded a $390 million deal to purchase 42.4% of the cobalt production from Vale's Voisey's Bay mine (Labrador, Canada) to help fund the mine's transition from an open pit to an underground operation. Voisey's Bay is one of the lowest-cost, highest-margin nickel mines globally, ranking in the bottom half of the nickel cost curve. Not only did this deal add a new revenue stream to the company, but it further strengthened Wheaton's long-time partnership with Vale, one of the largest diversified mining companies in the world.

A month later, Wheaton concluded a new $500 million streaming deal with Sibanye-Stillwater to purchase a percentage of its gold and palladium production from the Stillwater and East Boulder mines in Montana (United States).

Partnerships are a key part of the success of Wheaton. By providing sustainable funding options with no equity dilution, and ongoing support, both technically and from a community investment perspective, many companies often return with additional partnership opportunities down the line. On several occasions, Wheaton has partnered with the same company on multiple assets or expansions over time.

One of the advantages of the streaming industry is the partnerships that are created. The contracts are founded on a long-term relationship between two entities to create value for all stakeholders. This is compared to a royalty, which is just registration on land that says the miner must deliver a check to this person and they have a right to audit. For 15 years, I worked at mines with royalties covering them and you never saw the royalty guys except for once a year when they'd send in an auditor to test you.

Wheaton—The Preferred Streaming Partner

The management team at Wheaton is paid to grow the company but is also responsible for managing the existing partnerships so that maximum value is returned to all stakeholders. Programs such as the Technical Ambassador Program, where Wheaton supplies technical expertise at no cost, the partner Community Investment Program, where Wheaton co-funds projects around partner mining sites, and the Unlocking Value Program, where Wheaton is open to contract adjustments or amendments that unlock value for all stakeholders, all make Wheaton the preferred streaming partner, with many more repeat customers then any of its peers.

With Glencore, the company acquired two streams in Peru, almost a decade apart—between 2006 and 2015—for more than $1 billion. With Hudbay, Wheaton has streaming agreements on three assets that also total more than $1 billion.

That trend continued in 2021, when Wheaton and Capstone Copper entered into two precious metals purchase agreements in the same 12-month period over the latter's assets in Mexico and Chile. Wheaton had previous transactions with Capstone on the Cozamin (Mexico) and Minto (Yukon Territory, Canada) mines.

Haytham Hodaly, Wheaton's senior vice president of corporate development explained: "What we do is take the valuation that they are estimating for their ounces and show them the valuation we are estimating for ours. By doing this we can share that value spread to ensure that there is a win-win opportunity for both parties. The shared success on the first deal is often the catalyst to build a long-term relationship for future deals."

Deals can be done quickly too. While the company typically earmarks 8 to 12 weeks for a transaction, Vale wanted its $1.9 billion deal, which covered gold streams from the Salobo mine in Brazil and Canada projects, in under five weeks, and Wheaton got it done.

"Being a good partner means having a willingness to consider adjusting contract details and timelines when necessary to ensure mutual success," added Hodaly.

Another differentiator between streamers like Wheaton and other mine finance providers is that these partnerships extend to local stakeholders as well. Through its Community Investment Program, the company works with its partners to co-fund social and environmental initiatives in the communities residing in the areas in which Wheaton and its mining partners operate. Wheaton endeavors to work collaboratively with mining partners to support their social license to operate. One of our overarching principles within Wheaton is "the stronger our partners are, the stronger we are." So, by supporting and strengthening our partners social license to operate, we will share in that success. And it is the right thing to do.

Wheaton's commitment to sustainability was further cemented in September 2019 when it joined the United Nations (UN)'s Global Compact. As a participant, Wheaton committed to voluntarily align its operations and strategy with the 10 universally accepted principles in the areas of human rights, labor, environment, and anti-corruption, and to take actions that support broader UN global goals, including the Sustainable Development Goals.

We strongly believe in promoting responsible mining practices and supporting long-term sustainable benefits in the communities where we and our partners live and operate. Wheaton's asset portfolio is based primarily on low-cost, long-life mines, so for us, sustainability is critical to the success of our business model.

In 2020, Wheaton announced the launch of a $5 million Community Support and Response Fund (CSR Fund) to support global efforts to combat the social and economic impact of the COVID-19 pandemic. The CSR Fund was designed to meet the immediate socioeconomic needs of the communities in which Wheaton and its mining partners operate. This fund was incremental to Wheaton's already active Community Investment Program, which currently provides support to more than 50 programs in multiple communities around the world.

Further to this, in 2022, Wheaton committed to achieving net zero carbon emissions by 2050, inclusive of scope three emissions from its partner's operations. Although we have been operating as carbon neutral since 2015, we weren't including any emissions from the operating mines that we invest into. We get the benefit of metal production from these mines, so we owe it to society to take responsibility and report our share of the emissions. Again, it is the right thing to do.

Streaming, Royalties, and Community Investment Programs

Traditionally, royalty owners and streamers left the responsibility for community re-investment with the mine operators, even though they receive benefits and value from these operations. In early 2013, Wheaton changed that, with its first co-funded community investments alongside Barrick, Vale, and Hudbay. By 2022, Wheaton had invested over $35 million into community programs that supported small business initiatives and improved health and education facilities.

I am particularly proud of the fact that, whenever I see our peers entering into new agreements, they now follow our lead and make commitments toward directly supporting the communities around the mine sites. Without this cohesive approach to ensure that those most impacted by our operations are seeing benefits, our industry would not survive.

In little more than a decade, Wheaton grew from a company streaming less than 100,000 gold equivalent ounces in that first transaction to one that now has a growth profile climbing close to 1 million gold equivalent ounces per year from a diversified portfolio of operating mines and development-stage projects. This has been achieved with minimum equity issuances, effectively using debt and ultimately operating cash flows to continue growing.

"We rely very little on equity to fund our growth. Since inception, we have used operating cash flow and debt to fund about 80% of our growth. We haven't issued equity since 2016, and our current portfolio is generating about a billion dollars of operating cash flow a year," said Wheaton's Chief Financial Officer Gary Brown.

While it takes larger deals to move the dial, and more players are obviously participating in the market now than at the start, Wheaton has still managed to maintain a dominant position, having participated in three of the five largest streaming deals completed in the past five years.[12] These include the Salobo 1 and 2 deals, as well as that of Antamina. And through 2021 and early 2022, Wheaton completed nine smaller deals on development projects, with a cumulative value of over $1.5 billion. These assets will add an additional 20% to Wheaton's production as the mines come into production over the next two to five years.

Having now invested over $10 billion in streams since its inception in 2004, the company has already generated over $8 billion of free cash flow, and paid dividends of over $1.5 billion back to shareholders. And its current portfolio has 40 years of reserves and measured and indicated resources, and close to 20 years of inferred resources, spread over 32 assets.

When I think back to that boardroom meeting with Telfer in early 2004 and look at the value created from that spark of an idea, it is a bit overwhelming. And we just keep on growing, grounded in our vision to be the world's premier precious metals investment vehicle.

---

12. Wheaton Precious Metals, *Investor Day 2021: The Next Decade of Streaming* (September 22, 2021), www.wheatonpm.com.

5

ALTIUS MINERALS CORPORATION

TSX: ALS   OTCQX: ATUSF

www.altiusminerals.com

Brian Dalton

Altius was not actually set up to become a royalty company. The idea to form the company was put together by myself and fellow Memorial University of Newfoundland student Roland Butler in 1996 in the aftermath of the Voisey's Bay deposit discovery in our home province of Newfoundland and Labrador (Canada). We shared a common background in prospecting and optioning out projects to help fund our geology studies and had both benefited greatly when the Voisey's Bay area-play claims rush occurred. From this collective experience, we decided to form our own junior exploration company and brought in veteran geologist Geoff Thurlow and mining specialist lawyer John Baker to round out the team. Altius went public on the old Alberta Stock Exchange on Halloween 1997—two months shy of my 25th birthday.

Looking back, our timing could perhaps not have been worse nor better, depending on one's perspective. I vividly recall a meeting in Calgary with the sponsoring brokers as the IPO was being assembled during which other brokers from the firm literally burst into the meeting, holding a press release printout and asking us "Do you think there is any gold there?" They were referring to the latest news out of Bre-X Minerals, which was in the process of being exposed as one of the biggest mining investment frauds of all time. To say that this event put a sudden damper on the appetite of investors to fund the broader junior mining sector would be a massive understatement!

We did manage to complete the Can$300,000 IPO raise at 20 cents a share, but it was far from clear out of the gate when or if any more capital was going to be available to us. I'll never forget the first day of trading for Altius. We watched the screen until moments before the closing bell when to our great relief our first and only trade for the day finally went through. This was followed by a call from my parents to congratulate us on the listing—and with it an admission to my sudden suspicion that they had bought the only shares traded on our opening day as a public company.

The market didn't become more receptive to our enthusiasm over the coming months (though Mom and Dad would probably have mortgaged the house if I'd let them), and it began to sink in that our strategy of progressively raising money for the projects we'd assembled, and making a discovery in short order, was on rather shaky ground. Our response, borne out of fear, was to go back to our own basics. We decided to turn Altius into a slightly glorified version of our old independent prospecting ventures, which was to keep project-generation costs as low as possible and to try to attract partners to farm in and fund the more expensive exploration and drilling.

The logic was rooted in fending off corporate-level equity dilution in favor of taking it at the individual projects level. We accepted that this would mean less exposure to success on any one project while having many more shots on goal and a higher overall probability of discovery participation. A key element to this strategy involved a strong focus on retained royalties. While we did not consider ourselves a royalty company in the making at this point, in essence, that was indeed what was happening.

During 1998, we went on to deal five projects from our starting portfolio to partners that included BHP Billiton and Teck Resources Limited. This success brought in some modest but very welcome up-front cash payments, a future flow of news for the market and demonstrated that we could generate quality projects. It also caught the attention of a select few long-term junior mining investors: John Tognetti, Rick Rule, and Alan Yeung. These three individuals kept us going with a few hundred thousand dollars a year for low-cost project generation for the next few years while also being very generous with encouragement, advice, and mentorship that greatly enhanced our conviction to the business strategy.

We continued to add the likes of Barrick Gold Corporation, Inmet Mining Corporation, and Agnico Eagle Mines Limited to our list of partners. During this period, the broader public markets stayed absent, and an increasing number of our peer junior mining companies threw in the towel and changed their public vehicles into dot-coms, which had rapidly and dramatically emerged, at the expense of mineral exploration, as the new place to be for speculative investors. It really was that bad.

And it really was that good. Competition on the ground was essentially non-existent, and high-quality exploration lands could be assembled readily and cheaply by anyone with the conviction to believe that rumors of the death of the industry were being exaggerated. We loaded up.

Eventually, around 2001, and immediately following the dot-com crash, the mining market tide began to turn upward—or at least stop falling. We were early beneficiaries of this because we were not scrambling to re-convert our businesses back to mining but instead were flush with projects and partner-funded exploration news flow. Our share price crossed the dollar milestone, and we were being offered equity investment from many new quarters. Around this time as well, the term *prospect generator* (PG) began to gain broader adoption among industry pundits as an odds-beating strategy for exploration, and we were regularly cited as one of its strongest adherents.

We had managed to keep our capital structure tight and could even lay claim to being part of the creation of a new business model for the sector. We'd made it through.

What happened next would later prove to mark a very important point in our history. I took a surprise call from old prospecting buddy, Chris Verbiski, one of the original co-discoverers of the Voisey's Bay nickel deposit. He was touching base to gain some contacts to those in the financial part of the business who might be able to support him in selling some of his retained royalty. I took the question away and told him I'd give it some thought.

Within a day, I was back to him and said his first call was the only one he needed to make: Altius would like to buy the royalty interest.

It took a few months to sort out the terms (we both had to teach each other what net present value and discount rates meant), but we ended up negotiating to buy 10% of Verbiski's and Al Chislett's 3% Voisey's Bay net smelter return (NSR) for around Can$10 million in cash plus some equity warrants, which was money we needed to raise. As I recall, our market value leading into that financing in early 2003 was around Can$13 million, so the decision was not a light one given the equity dilution implied.

The rationalization for the move, however, was still not based in transforming our business into a royalty company. Instead, we saw it as the last sale of shares that we would ever have to undertake and a means of jumping off the equity funding treadmill. We estimated that our future annual revenue from the royalty would be sufficient to fund our ongoing project-generation activities and corporate overheads for as far into the future as we were capable of envisioning.

> ## Long resource lives are the greatest predictors of future upside.

Fast-forward a few years and the strongest mining market anyone alive could remember was underway. An incredible surge in metal demand out of China and other BRIC countries (Brazil, Russia, India, China) provided the backdrop for increasing references not just to a normal mining bull market cycle but indeed a supercycle that could endure for ages to come. So the story became: "It was different this time."

Voisey's Bay had come into production and was paying us based on nickel prices that were multiples of what they were when we bought the royalty. We now also had insatiable demand for the projects we were advancing, and the value of these shot up dramatically. Practically any junior mining promoter had easy access to capital. We dealt out a great many projects and added a new twist to our strategy: We began to spin out projects into new public companies for large equity positions and retained royalties. Keeping on the fast-forward button, within a few more years, we had turned several of these equity positions into more than Can$200 million in monetized capital gains versus original exploration spending of somewhere around Can$10 million. It was one hot market!

Compared to our Can$300,000 IPO raise, this amount of cash was mind-boggling to say the least. However, it came with its own new challenges. How could we deploy this much money as a project generator whose success we believed was heavily attributable to keeping front-end generation costs as low as possible? Would an individual success on any project, be it a new deal or even an average scale exploration discovery, move the needle anymore with that much cash sitting on the balance sheet?

We had a few ideas, with an early favored one being to simply return the bulk of the cash to shareholders to shrink the base value of the company back to the point that our project-generation business successes would be relevant again. Polling of shareholders on this idea, however, did not meet with the enthusiasm we expected. Anyone invested in mining at this point in the cycle was generally flush with cash already, and so we were challenged (in some cases almost angrily) to reinvest and further grow the business. We went back to the drawing board.

What we knew we would not do, in spite of many incoming proposals, was try to evolve into a mining company. We had enough sense to know that the skillsets that had gotten us to this point did not naturally transfer to running a complicated industrial business. The idea in fact terrified us.

As to royalties, Voisey's Bay had exceeded our wildest hopes. Payback had come in a fraction of the time we had originally envisioned, and this had obvious emboldening impacts. Moreover, by now, we had a sizable portfolio of earlier-stage royalties because of all of the projects we had optioned to other companies and retained royalty interests against. The strategy we landed on was to use the cash to acquire more advanced-stage and cash-flowing royalties to fill out a true pipeline portfolio. We were now finally admitting to ourselves that our destiny was as a royalty company—albeit one with a continuing exploration-based growth strategy differentiator.

We further decided that our focus area would not be the precious metals part of the mining royalty world. There were two key reasons for this. The first was that there were already some very formidable players active in this space whose competitive strengths are well described in other chapters of this book. The second was that we had a view that base metal and bulk commodity mines had inherent features that made them, on average, more upside laden than the average precious metal mine. The difference related to the greater operational and capital scale that was involved. Many gold mines could be built around fairly modest resources and implied mine lives and a relatively low capital investment amount. Base and bulk mines generally require larger resource lives to justify the greater capital required to build them, as well as bigger, stronger operators. We simply believed that these factors would translate into more option value in the form of more opportunities for operator-funded life extensions and output expansions across multiple cycles.

We made a wish list of projects that we would like to hold royalties on and went to work to try to entice the owners of existing royalties to sell, and/or convince project operators to create new royalties. The challenge we quickly ran up against related to the same strong market conditions that had allowed us to build our war chest to begin with. Existing operators had many attractive alternatives from the equity and debt markets for any new capital requirements, and existing holders had stars in their eyes because underlying commodity markets were so strong and were being projected to remain so essentially forever under the "different this time" narrative. Our memories of the difficult markets we had started in remained strong, however, so we decided to wait for a better—really meaning a poorer—market backdrop.

The wait turned out to be a long one. From when we began to communicate our strategy of directly buying more advanced-stage royalties to the first major acquisition took more than six years! During this wait, we did stay busy with our project-generation activities. After all, our new focus on royalty buying was seen as a business line addition rather than a full change of business.

We were not abandoning the golden goose but just trying to buy some more eggs to complement its ongoing output.

We also found occasional opportunity along the way in buying equity in other royalty companies whenever the market volatility priced them lower than we believed was the value of their underlying royalties. A notable example of this was International Royalty Corporation (IRC), which by then held the remaining part of the Voisey's Bay royalty that we didn't own, as well as a portfolio of other royalties that covered a spectrum of precious and non-precious assets. Over time, we became a significant shareholder of IRC and at one point even approached it with an idea of merging and spinning out the precious metal royalties. This was not to be, however, as some of those other precious metal royalties it had assembled were meaningful enough to capture the interest of precious metal royalty heavyweights Franco-Nevada Corporation and Royal Gold. They both ended up bidding for IRC, and we eventually signed a lockup with Royal and tendered our shares to their bid. We made a fantastic return but still had no new royalties added to our portfolio. Two other investments in royalty companies that we made during the wait included Callinan Royalties Corporation and Virginia Gold Mines. More on those later.

At the top of our earlier mentioned royalty target list was a portfolio of royalties held by mining company Sherritt International Corporation that related to most of the major Saskatchewan potash mines. In terms of mine lives (hundreds to thousands of years) and related optionality, these are unrivaled globally. We tried to buy them several times, but the answer was always no, and we even were told at one point to stop bugging them because these will never be for sale. However, in late 2013, with mining markets now in freefall, they suddenly were.

Sherritt was under balance sheet pressure related to a big nickel mining project development in Madagascar, and so it began a process to package sale its potash royalty portfolio, as well as a collection of Alberta-based thermal coal royalties. The ticket size upon packaging was too big for us as our early appraisal suggested it would transact for somewhere north of Can$500 million against our market cap at the time of roughly Can$300 million. It was clear that we would have to partner up and also leverage what we had available as a cash equity contribution to be taken seriously in the sales process.

Through a relationship we'd built on the PG side, we had an open invite to present deal ideas to Liberty Mutual Group, a major U.S.-based insurance company, whose mining investment team was led by Noel Dunn, now of Ero Copper Corporation. Dunn immediately recognized the annuity-like characteristics of the potash portfolio, and Liberty agreed to become our bidding partner for a half interest.

In parallel, we began discussions with Scotiabank around arranging a senior lending facility to support our share of the bid. We had never borrowed to this point and so had no credit history, but we did have someone senior at the bank who was willing to vouch for us. I had been asked a few years previously by another mentor, "Commander" Harry Steele, to join a trip to Labrador to perform a specific mission. Steele had said, "Dalton, we got a fella coming up to the camps who's gonna be a big deal in Scotiabank and you are going to show him how to catch Labrador salmon." In hindsight, it was actually more of an order than a request.

Brian Porter, then a senior executive and later CEO of the bank, turned out to be a keen and enthusiastic student. One day, the Commander sent us off in his helicopter to fish the remote and magical Falls Pool on the Lewis River, and the lesson for the day was on properly drifting and twiggling dry (surface) flies. These flies are tricky to fish for salmon but when worked right can produce highly visual and incredibly exhilarating takes from this mighty fish. The 10-pounder that eventually exploded on Porter's Green Bomber fly, after several pulse-generating false rises, was a star performer and Atlantic salmon fly-fishing instantly claimed its newest addict. And I had a great new relationship started for whenever the day came that we needed the support of a major bank.

The bidding process proved tough and very competitive, but in the end, we prevailed and suddenly went from Can$3 million to Can$28 million in royalty revenue and from 1 to 13 underlying mine exposures. We'd finally made good on our promise and became a leading diversified mining royalty company. This was mid-2014.

The mining market was firmly back in horrible territory at this point. Prices just kept dropping on expectations of supply surpluses and falling demand. A lot of new mine construction had been incentivized during the preceding bull market, and production from many was now in sight. At the same time, the "strong demand growth forever" narrative from the BRIC countries was breaking down. Moreover, much of that new construction was heavily funded with debt, and the projected operating margins were starting to look thin against the low prices. It turns out that even supercycles have two sides to them, and this was becoming a super-bear indeed. As 2015 advanced, the negative sentiment was reaching dull roar levels, and the entire mining industry was suddenly placed on collective credit watch. Balance sheets that had earlier looked rock solid were now evolving into every chief financial officer's worst nightmare. Asset sales and monetizations were very much on the table, and the royalty and streaming sector was being heavily called upon to write checks to big and small mining companies alike.

> *The prospect generator is the golden goose and the royalties are its eggs.*

Unfortunately, we were now fully tapped out on liquidity after the Sherritt deal while conditions continued to improve (or deteriorate depending once again on your perspective). There were still targets on our wish list, and they were looking likely to come to market. We had to do something to get back in the game.

The first bit of liquidity relief came early in 2015 when one of our royalty company investment holdings was taken over. We had become major shareholders of Virginia Gold after it was spun out as a royalty vestige of Virginia Mining upon its takeover by Goldcorp (now Newmont). Virginia Gold, led by industry superstar André Gaumond, held a retained NSR on the development-stage Eléonore gold mine in Québec (Canada) and it agreed to be acquired by Sean Roosen–led Osisko Gold Royalties, allowing Altius to turn a meaningful profit and improve its balance sheet by paying down some of its debt.

We were also at that time still a significant shareholder of Callinan Royalties, which held a cash-flowing royalty on the Hudbay Minerals 777 copper-zinc-gold mine in Manitoba (Canada) and also had a nice slug of cash in the bank. It further held a portfolio of earlier-stage equity and royalty interests that had been put together by former Altius co-founder Butler. Butler's genius for spotting exploration opportunity early borders on the supernatural, and so we knew inherently that there were hidden gems in there.

Prior to that point, we had been somewhat deterred from approaching Callinan from a mergers and acquisition perspective because its 777 royalty was structured as a net proceeds interest (NPI) product compared to the revenue-based royalties we strongly preferred. When Callinan announced that it had negotiated a deal with the mine operator to convert the NPI into an NSR, this impediment fell away, and we began to negotiate a merger. For Callinan shareholders, the benefits were in gaining exposure to a larger and more diversified portfolio of royalties versus the relatively short-life single asset exposure that 777 offered. For us, the primary appeal lay in adding direct cash while gaining several years of strong cash flows from a strong mine to match up against our debt repayment schedule, the combined impact of which was to dramatically deleverage and de-risk our balance sheet. We had regained liquidity and were back in the market as buyers.

All around us the precious metal royalty players were having a field day by buying gold and silver by-product credits from major base metal miners. Our move was to try to find the flip side scenario: gold miners with significant long-term base metal by-products. In early 2016, essentially at what later proved to be the market bottom, we announced the acquisition of a copper stream from Yamana Gold relating to their Chapada copper-gold porphyry mine in Brazil. While this brought us an attractive new long-term revenue stream based on the existing mine plan, what really caught our attention was the future promise of the geology, recognized excitedly by the geologists within our PG team, to host far more ore than was within the current plan. We felt strongly that Chapada would run for much longer and potentially at higher production rates than the market was giving it credit for. We feel even stronger about that today.

The last thing we invested in during this wonderful window of opportunity was a starting position in shares of Labrador Iron Ore Royalty Corporation (LIORC), a former income trust that now served essentially as a corporate pass-through vehicle for royalty and dividend income related to the long-standing operations of the Iron Ore Company of Canada in Labrador. Benchmark iron ore prices briefly traded below $50 per metric ton at the cycle bottom, and we believed that the underlying royalties held by LIORC were intrinsically worth far more on a long-term basis than the capitalization being assigned to the company in that gloomy market backdrop.

As 2016 went on, the balance sheet crisis in mining ultimately faded and market sentiment and mined commodity prices began to rebound. The trigger seemed to be promises from newly elected U.S. president Donald Trump to make massive investments in infrastructure, with obvious implications for metals demand. This improvement in markets caused the distressed asset selling window to largely dry up, but we were happy with what we'd managed to pick up along the way, feeling we'd made the most of it. Our portfolio now consisted of 15 royalties generating Can$60 million in revenue.

The other thing we started to see emerge that year with the better market sentiment was a renewed interest in exploration lands. We began to deal projects from the big portfolio of properties we'd managed to assemble through the downturn, quickly converting several of these into new junior equity positions and early stage royalties.

We suffered a major blow that November, however, when news broke that the new Alberta New Democratic Party government had reached an agreement to pay the operators of coal-power-generating plants, several of which were paying us royalties, to phase out their operations by 2030. This had the effect of truncating the life of our Genesee mine (Alberta, Canada) royalty by some 25 years, and we took a significant write-down accordingly. While there is no sugar coating the negative impact of this effective expropriation, in Alberta of all places, we did eventually find a way to make lemonade. Because of that blow, and our desire to find a long-term replacement for the future lost revenue stream, we began to explore the royalty possibilities around what was emerging globally as the replacement for coal-fired electricity generation, namely, renewable energy.

Before getting into this, it is worth highlighting an unexpected meeting that took place to close out 2016. One of Altius's directors, Fred Mifflin, happened to know Fairfax Financial Holdings Limited founder Prem Watsa on a personal level, and he asked me if I'd be interested in chatting with him. They had been talking about the state of Canadian entrepreneurship, and Mifflin had brought up Altius. Mifflin also knew that I was an admirer of Watsa, who is often referred to as Canada's Warren Buffett. There was no business agenda involved, simply an opportunity to meet a legend—akin in my mind to being offered a chance to meet Wayne Gretzky (whom incidentally Watsa later did introduce to me)!

We got together a few days before Christmas, and it was clear from the outset that he was more interested in hearing about our history and philosophy and the trials and tribulations of our journey, than about the current status of the business. Looking back, I don't think we spoke in any kind of detail whatsoever about the assets in the company or its revenue profile. This added to the surprise when at the end of our chat, he asked me what we would do if we had more permanent capital to work with—about Can$100 million more. I mumbled and stumbled my way out of the room saying I'd have to get back to him on that.

A month or so later, we were back detailing some ideas. These included the value proposition we still saw in the shares of LIORC. We explained that, due to its pass-through structure, buying its shares was akin to buying a long-life, time-tested royalty on a crown jewel Canadian mining asset. We also brought up the idea we were working on to innovate the royalty model into the renewable energy sector.

Soon after, we closed a Can$100 million, 5% preferred share financing with Fairfax that had an almost 100-year maturity, which was not quite permanent but tied into one of Watsa's favorite sayings, "The first 100 years are the hardest!" Before long, we declared ourselves as a major shareholder of LIORC and announced that we were partnering with an expert team based in New Hampshire to develop a renewable energy royalty business.

The third investment made with that funding, leveraged with more term borrowing led by Scotiabank and several other international banks, was deployed at the beginning of 2018 when we bought Liberty Mutual out of the interest they had acquired alongside us in the Saskatchewan potash royalties. Dunn had since moved on, and Liberty wanted to exit its mining investments. Potash prices had done okay since the original acquisition, and the royalties were performing well. We, therefore, happily paid a small premium over the original purchase price given our longer-term view on the even greater remaining embedded upside.

Since then, mining and commodity markets have done better (with some notable volatility along the way coming from trade wars, a pandemic, and now runaway inflation and a war in Europe). Because sustainability-focused investments have the potential to drive increased metals demand, there is renewed talk of another supercycle in the making. Our revenues have continued to grow with prices and volume improvements, and today we are generating more in royalty revenue *per day* than we netted in our IPO. The yields against our purchase prices during the downturn have also shaped up fantastically. In fact, paybacks have been already reached or are approaching for most of our royalty purchases, while our remaining average resource lives arguably lead the mining royalty sector. More importantly perhaps, mining companies are making strong margins and have finally begun to at least try to gain courage around investments in mine extensions and expansions and even new builds. We believe that as this investment eventually goes to ground, we will be long-term beneficiaries, at no incremental cost, given that so many of our investments were specifically selected for such an eventuality.

Our PG business is also enjoying the better conditions and generating healthy capital gains for us while continuing to steadily add to our early stage royalty portfolio. Time is also becoming our friend with respect to the exploration royalties we began creating 25 years ago, most of which have minuscule or even negative cost basis. Note that 25 years is considered around the average time it takes for a new discovery to reach production in today's modern mining industry. We have some excellent candidates in the portfolio to potentially get built this cycle and, should that happen, the returns will be mathematically infinite and further our progress toward a long-standing differentiation goal of generating the strongest fundamental investment returns in the sector.

Altius Renewable Royalties (ARR) has been building strongly and has thus far acquired royalty interests in more than 30 U.S.-based wind and solar royalties that have a total combined generating capacity of more than 7,000 Megawatts. Some of the royalties have been direct purchases while others have stemmed through funding arrangements with developers, which are essentially the project generators of the renewables world. Corporately, ARR has formed a joint venture with major private equity firm Apollo Global Management and completed a Can$100 million IPO. The current market value of ARR to Altius shareholders, net of founding investments, has now also solidly eclipsed the write-down hit we took on Alberta's coal-based power generation phase out—and it is really just getting going.

The things that excite us most about the ARR opportunity are the sheer scale of the addressable market and that the natural resources we are creating royalties around will never run out or deplete. We continue to believe that long resource lives are the greatest predictors of future upside and optionality realization in the royalty world—and perpetual is a very long time indeed.

Before closing off, there is an interesting end note to share with regards to the Callinan merger deal. The 777 mine has just wound down as I write this, pretty much as per schedule at the time of our acquisition, but the overall investment continues to bear fruit. One of those "hidden gem" exploration interests we'd assumed Butler had envisioned and acquired at Callinan has been revealed as a royalty on a project called *Silicon* (Nevada, United States).

AngloGold Ashanti Limited has recently published a multi-million ounce maiden gold resource at Silicon while hinting at the discovery of a potential broader world-class cluster of new deposits within the royalty area. Altius, the avowed non-precious metals–focused royalty company, somehow finds itself with a royalty covering a major new Nevada gold district. We couldn't make this stuff up if we tried!

Our 25-year path to becoming a meaningful diversified natural resource royalty business has obviously not been straight or conventional. We evolved into it over time while maintaining our roots in project generation. In fact, when we look at our mining royalty portfolio, we see pretty much everything stemming originally from project generation. This includes not only the royalties this part of the business directly created, but also those that were bought by leveraging its profits. This is why we say that PG is the golden goose and the royalties are its eggs. It's perhaps not the easiest or fastest way to go about building a royalty company, but it is one way, and we are proud to say, it is our way.

6

INTERNATIONAL ROYALTY CORPORATION

TSX | AMEX: IRC

Douglas Silver

The story of International Royalty Corporation (IRC) began with my work as a mineral appraiser. I am a geologist who morphed into a mineral economist. I really enjoyed doing valuation work so I enrolled in the American Society of Farm Managers and Rural Appraisers and took their two-year course of study to become a certified rural real property appraiser. Because there were only about a half dozen certified mineral appraisers in the world, my consulting company, Balfour Holdings, was swamped with valuation work.

In 2001, I received an appraisal assignment from High Desert Mineral Resources, a small public royalty company with assets centered in Nevada's Carlin trend. At the time, 93% of the outstanding shares were owned by Lee Halavais, the founder. She decided to take the company private and contracted me to do the appraisal. About halfway through the process, Royal Gold reached out to Halavais with an offer to purchase her company. Because she had no experience in selling companies, she asked me if I could take on the responsibility. I switched my roll from being an appraiser to being a negotiator.

This led to an intensive nine-month process, learning everything there was to know about the mineral royalty business. I studied how royalty companies were structured, their business plans and investment objectives, and most importantly, how royalties were valued. I found the model fascinating because royalty companies were scalable and could be run on a shoestring with few employees: Buy or create the royalty, clip the coupon, and use these funds to search for the next opportunity. I preferred to do the targeting and analysis work myself, using the vast mineral deposit database that I had assembled for my consulting work, so the royalty model was perfect.

Additionally, while consulting offered a wonderful lifestyle and learning experiences, it really didn't make much money, and my two children were about to enter college. I also recognized the positive impact of the emerging China-driven mineral supercycle and knew it would be the perfect time to establish a public company and float a stock offering.

During this same period, I kept running into Doug Hurst, a fellow geologist turned mineral economist. We—the two Dougs—frequently gave presentations on the same conference panels, often invited because of our willingness to share our private research on the mineral industry, which was derived from the large proprietary databases each of us had developed. Several of these databases significantly overlapped, so we began to search for ways to collaborate, which ultimately led us to form IRC—in the process, the two Dougs began a lifelong friendship.

The third founder of IRC, George Young, was an American lawyer fluent in Spanish, who had worked with me at Bond International Gold. Young's ability to think out of the box on deal structures and his vast experience in Latin America added a third leg to the team.

Because Hurst was based in Canada and I was based in the United States, we each had extensive but separate networks to tap for opportunities. Our databases closely tracked royalties, so we unknowingly had created a comprehensive starting list of royalty acquisition targets.

Hurst knew Claude Bonhomme, who controlled a 0.25% net smelter return (NSR) royalty on the world-class Williams gold mine in Hemlo, Ontario (Canada). Bonhomme's royalty at the time paid out about $500,000 per year, which was a fortune in the eyes of Hurst and me. By purchasing the royalty, we would have the seed funds to allow us to close down our respective consulting practices and focus solely on building IRC.

The problem was how to pay for the royalties we identified. The obvious answer was to go public. Because Hurst had worked as an analyst in the Canadian investment banking sector, he arranged meetings for pitching the banks. But after a dozen or so such meetings, the only answer we received was no.

The fundamental problem was that the bankers didn't understand the royalty business. In 2002, Franco-Nevada Corporation and Royal Gold were the largest mineral royalty companies in existence, and there were fewer than a handful of public mineral royalty firms. Wheaton River Minerals (now Wheaton Precious Metals Corp.) was not considered a royalty company at that time because they were pitching themselves as a synthetic silver miner with a strange new business model called *metal streaming*. Additionally, the royalty companies focused only on precious metals. To the Dougs and IRC, a dollar of cash flow was worth a dollar of cash flow and the metal that generated it was irrelevant, so consolidating non-precious metals royalties gave us a competitive market advantage. Alas, we were wrong. Even today, the financial markets continue to think that a dollar of gold cash flow has more intrinsic value than a dollar of nickel cash flow. This logic bomb plagued IRC funding efforts for years.

> *This demonstrates how important strong, long-time business relationships are when building a company with no money.*

The Dougs maxed out our credit cards searching for an investment bank that could recognize the value premise of a multi-commodity royalty company and had the capability to take IRC public. We had one meeting left; we arranged to see John Tognetti and John Rybinski of Haywood Securities in Vancouver, Canada. I was so convinced that we would be rejected that I had set up a meeting with one of my Vancouver-based clients an hour afterward so I could pay for my flight home.

The meeting began, we Dougs made our case, and around minute 15 of a 30-minute presentation, Tognetti said, "Stop." The disappointment was palpable for about two seconds until he followed up with, "I will take the whole deal." The problem was how to lock down the deal with me needing to leave the meeting to restart my consulting practice! The quick solution was to leave Hurst to work through the details with the Haywood execs. Hurst's knowledge of IPOs and private placements much exceeded mine at the time, so he was the perfect person to leave behind.

The support by Haywood laid the groundwork for launching IRC. They provided an initial private placement of Can$3.52 million in 2003. With money in the bank, IRC was now able to close the Williams transaction and start looking for other deals.

Assembling a portfolio of diversified royalties was feasible at this time because we could approach our mining clients and ask to buy their royalty portfolios. Most had long-known both of us professionally and understood IRC didn't yet have the level of financing required to pay for these royalty assets. But they were willing to wait on proceeds from a successful IPO to get paid. This demonstrates how important strong, long-time business relationships are when building a company with no money.

To establish itself as a real royalty sector player, IRC needed a flagship royalty the same way Franco-Nevada had its Goldstrike mine (Nevada, United States) gold royalty and Royal Gold had its marquee gold royalty on the Pipeline mine (Nevada, United States). The opportunity to accomplish this came in the form of a unique circumstance to acquire a significant portion of the royalty attached to the massive Voisey's Bay nickel project in northern Labrador, Canada.

Voisey's Bay is arguably one of the best nickel mines in the world. It was owned by Inco Limited, one of the world's largest nickel producers at this time (Inco was subsequently acquired in 2006 by Vale). The deposit contained extremely high-grade nickel and copper ore and was minable via an open pit. For those in the gold business, the grades at Voisey's Bay were equivalent to having an open pit gold mine with head grades of 3 ounces per metric ton.

Newfoundland native Chris Verbiski is a self-made man, having taught himself geology and practical mineral exploration. Along with his partner, Al Chislett, they discovered the Voisey's Bay deposit in September 1993. The partners received a 3% NSR on the deposit as part of their success fee. They sold 10% of this royalty to Altius Minerals Corporation, whose CEO was a fellow Newfoundlander. This transaction suggested that the other 90% of the royalty could potentially be acquired as well. The critical issue was the simple question of price.

Unfortunately, Verbiski was one of the very few industry players neither of the Dougs knew. I reached out and asked Verbiski what his favorite hobby was. "Fly fishing," he responded. This was an amazing coincidence as both Hurst and I were also enthusiastic fly fishers. The only problem was that Verbiski is one of the world's best fly fishermen; comparably, the Dougs are hacks!

A three-day fishing trip in Newfoundland ensued so that everyone could get to know each other. Verbiski brought his chief financial officer (CFO), Chris Daly. It rained most of the trip, but fish were indeed landed and discussions initiated between the Dougs and the Chrises. This began a process that took nearly a year for Verbiski and Chislett to become comfortable with IRC and allow it to land the remaining 90% of the NSR. The price tag was Can$160 million. Prior to this deal, the highest amount ever paid for a mineral royalty was $36 million. This focused significant investor attention on IRC.

While establishing its initial portfolio of royalties, IRC made a bid to BHP Billiton to purchase its royalty on the world-class Oyu Tolgoi copper mine in Mongolia. My long historical relationships with BHP managers made this possible. Ivanhoe Mines, however, had a right of first refusal on this royalty and exercised it in about 1/10 of a microsecond after being notified, paying $37 million. A frequent question asked by potential investors during the IPO road show was "What other royalties have you looked at?" We would relate the Oyu Tolgoi story and apparently, they were impressed. We quickly learned that swinging for the long ball, even when you are unsuccessful, will often build your credentials among investors.

By 2005, the team had purchase-option agreements in place for a portfolio of more than 60 royalties with Voisey's Bay as the flagship. On February 23, 2005, IRC closed on Can$190 million financing, the largest mining IPO in Canada that year. The listing was successfully accomplished on the Toronto Stock Exchange as TSX: IRC.

The IPO launch deserves additional commentary, as it all hinged around simultaneous closing of the Voisey's Bay and BHP deals. But this was a serious catch-22 situation—closing on the royalties couldn't happen unless we had the money to pay for them! Furthermore, the IPO structure was complex as it involved both Can$30 million in debt and Can$160 million in an equity raise.

Consequently, the closing was divided into three geographic locations, each in its own time zone. Young went to Vancouver to complete the BHP portfolio acquisition, I was in Toronto to close the other portfolios, debt, and equity portions, while Hurst headed to St. Johns to close the Voisey's Bay transaction. For almost 72 hours prior to closing, it is doubtful that any of us slept. Toronto-based lawyers Mike Bourassa and Nancy Eastman of Fasken Martineau DuMoulin served as IRC's lead attorneys and had more than 20 or 30 lawyers assisting them, finalizing the voluminous paperwork. Each royalty had its own contractual agreements and attachments, and all of the documents had to be executed in triplicate. It took two giant conference rooms at Fasken's headquarters just to hold the papers.

Anyone who has worked with debt financing knows well that the paperwork is endless. Furthermore, the debt lawyers started squabbling over minute details, which of course kept setting back the clock. Without the debt piece, everything would be lost. The enormity of the ordeal—getting so many documents processed and executed—became stressful in the extreme, especially on so little sleep.

Mark Wellings, an investment banker with GMP (an IPO co-sponsor along with Haywood Securities), stepped up to the plate and took charge like a Marvel superhero. He told the debt lawyers for the lender that if they didn't get their act together immediately, no one was going to get their fees paid because the deal would be called off. This focused their attention, and signatures quickly followed.

The money was now in place allowing the Voisey's Bay deal to close in St. Johns. But the phones didn't ring in Toronto. The Toronto team waited to hear, but no one was checking in. After about an hour, Hurst at last called me.

"Doug," he said, "We have a problem."

"How big?"

Hurst: "Huge."

"Please explain."

Hurst: "We have run out of documents to sign!"

With that, Hurst will forever be the recipient of my revenge pranks. When Young closed the BHP portfolio at the same time, IRC was officially launched!

With money in the bank and this important set of royalties under our belt, the next five years were spent growing our asset portfolio. Each royalty we acquired is worthy of a story of its own because each purchase opportunity invariably developed its own personality as it moved through the process. There are immense differences in how one negotiates with a corporation versus an individual. Detailed technical understanding of the geology, resource, mining conditions, metallurgy, commodity markets, permitting rules, and economics associated with any mineral asset is paramount. Tax treaties must be reviewed, and tax planning is critical when cross-border transactions are involved. Regulatory approvals concerning title registrations must be studied. And then there are the personalities of each individual seller. It is particularly important to have a legal firm with multi-faceted expertise in these subjects, and our Fasken team members proved their value many times over.

The initial board of directors consisted of myself (chairman), Hurst (director), Young (director), Rene Carrier (lead director, former broker, policeman, and compliance expert), Daly (representing Verbiski/Chislett and also an accounting expert), Colm St. Roche Seviour (representing Verbiski/Chislett and a highly regarded Newfoundland attorney), Ed Mercaldo (independent director recruited at the recommendation of Haywood as he had served as the investment banker involved in the sale by Diamond Fields Resources of the Voisey's Bay project to Inco), Gordon Fretwell (independent director and Canadian securities lawyer), and Robert Schafer (independent director). Our directors were invested in making IRC a success, and each generously contributed their expertise to the IRC effort. Having a solid board provided strong oversight and support to senior management.

The core management team consisted of me (CEO), Hurst (president), Young (vice president and general counsel), a small accounting team (Ray Jenner, CFO; Brian Funkhouser; and Yelena Ovanesyan), and a large business development team: Jim Lydic, (vice president of business development), David Hammond (vice president of strategic planning), and Jim Jensen and Art Courtney (business development managers). Jack Perkins ran the investor relations program and Tu Li was both the office manager and my personal assistant. Paul Zink later joined the IRC team as its president.

Here is a typical example of what transpired when we tried to purchase a major mining company's royalty portfolio. IRC's databases were comprehensive after having been created and maintained for 20 years. We would approach the mining company to ask if they would consider selling their mineral royalties, most of which had evolved from exploration property transactions. In our initial overture, the mining company was provided with a list of their royalties extracted from our databases to see if we were missing anything. More than once, the company would respond, "We don't own that particular royalty." We would then have to show them that yes, they indeed did own it. A common reaction was, "Our land department hates you because you identified assets they were supposed to know about." This was hardly our intention, but it did bring up an interesting characteristic of land management departments, especially in large organizations during this period.

This situation repeatedly occurs because most royalties are formed at the exploration-stage of a property's life, where the seller would retain a small economic interest, in the form of a royalty, just in case a valuable ore deposit is proved up by a successor owner. In practice, these retained royalties are almost an afterthought. Mineral exploration is a high-risk business, and success is an elusive goal. Exploration properties frequently get dropped, so many of these royalties die a natural death, yet remain on the dusty rolls of the land departments. It is not until a discovery is made and a resource is defined that mining executives focus on the royalty clause of exploration agreements.

Sellers often kept royalties when they sold an asset to hedge their bets. What if the property turned out to have greater value down the road? A discovery could be made, a resource defined, or a new mine sent into production. By holding a royalty, they could recapture some or all of this possible future value.

With the emergence of a formal royalty industry, sellers knew they could get another bite at the apple by retaining a royalty, with the possibility of selling it to a royalty company. The gradual transitions to include streaming structures widened the opportunity set. Many of the individual royalties and portfolios pursued by IRC followed this process template. It is doubtful, however, that this approach could be productively employed today. After 30 years of royalty companies picking through the world's royalty endowment, existing portfolios of any merit have effectively been cleaned out. Today, mineral companies deliberately create new royalties for the sole purpose of auctioning them off to the much larger field of potential royalty buyers.

Over the years, we identified and analyzed royalty opportunities across a wide spectrum of mineral commodities, including but not limited to base and precious metals, iron ore, nickel, potash, coal, frac sand, and uranium. Our evaluation process had to be highly structured because the global mining industry was aware of IRC's business strategy, its near monopoly on non-precious royalties, and our financial capacity. We saw increasing numbers of royalty purchase and/or creation opportunities rather than having to rely on our databases. After a quick initial screening to sort out the deals that met our criteria, opportunities were typically assigned to one of the IRC officers, who would be responsible for all aspects of the potential transaction. This would involve the detailed technical investigation, with a focus on geology, reserves and resources, on-site and downstream processing, markets, and product pricing. Detailed economic evaluation and risk assessment would follow. After moving through the next level of review by the executive team, legal and land investigations would proceed, as well as due diligence assessment of the target's owner/operator and its management team. The latter often became the most critical factor in successfully creating a value-adding royalty, and our team paid increasing attention to this component. The lead officer would present the package and recommendation to the team for an approval decision. We made it a rule that the approval had to be unanimous because each of us looked at the project differently. If one member had an issue, we dug deeper to see if this issue could be resolved. Approved projects next moved to the board of directors level for final review and sign-off. The lead then conducted the negotiations with the seller, worked with outside technical consultants and legal counsel to perform more thorough due diligence, and prepared the closing documents.

Naturally, many investment opportunities are discarded along the way for a host of reasons. In later years, IRC would typically run through more than 150 possibilities annually. Locations were all over the world, from diamonds in Northern Canada, to iron ore in Western Australia, to gold in Burkina Faso, and copper in Brazil and Chile.

Mineral exploration is a high-risk business, and success is an elusive goal.

Every business enterprise needs an exit plan, even it is merely to indefinitely continue as a going concern. After five years of hard work, the Dougs, the rest of the management team, and the board were certainly looking at such continuation. But the times, as they say, were "a-changing." At the CIBC bank annual dinner in January 2009, David Harquail, CEO of Franco-Nevada, pulled me aside and posed the following: "Doug, Franco would like to purchase IRC. We know your company well, but there are two areas that we need information on. Would you consider letting us sign a limited NDA [nondisclosure agreement] to study these two issues in order to prepare an offer?" The two issues had to do with the transfer-pricing challenges at the Voisey's Bay mine, which IRC was intensely involved with efforts to resolve, and a tax-shelter deal IRC had made with McWatters Mining.

About a week later, I received an unsolicited call from Tony Jensen, CEO of Royal Gold. It went as follows: "Doug, Royal Gold would like to purchase IRC. We know your company well, but there are two areas that we need information on. Would you consider having us sign a limited NDA to study these two issues in order to prepare an offer?" It was virtually the identical call, and their concerns were with the same two issues.

We may not be superstitious but having two of the world's largest royalty companies contact us in such a short period with the same questions told us that IRC was officially in play (whether we liked it or not).

Large transactions tend to have project names in order to disguise their identity. Our investment banker, Scotia Capital (now called *Scotiabank*), generously allowed me to call our corporate sale "Project Titanomachy." Titanomachy means "the battle of the Titans."

By January 2009, both the mining industry and the mineral royalty business had begun to change. Most of the existing flagship-scale royalties had been acquired by the public royalty companies. There were also fewer mining company portfolios of exploration and development royalties available for sale. But to stay relevant to institutional investors, a royalty firm was expected to grow its revenues by at least 10% per year. The royalty companies had achieved much larger market capitalizations by 2009, so finding royalties large enough to fulfill this mandate became increasingly difficult. Thus began the transition from a royalty-based to a metal streaming–based industry.

IRC had a short list of eight opportunities it had targeted and was actively pursuing; it needed to land at least one of these to remain relevant to the investing community. Simultaneously, then U.S. President Obama was threatening to raise the estate tax, which brought into question why anyone would work 60 to 80 hours each week only to give most of that money to the government. Last, there was executive fatigue. The Dougs, along with their new president, Paul Zink, were continually on the road meeting with existing and potential new investors. It was estimated that we three collectively held more than 125 one-on-one meetings every year. Most people think of travel as glamorous and enjoyable; however, when a person is on the road for upward to 100 days per year, it becomes grueling, especially in times when the share price isn't responding to the hard work put in.

With the global market crash of 2008–2009, IRC's share price collapsed from Can$8.90 to Can$1.32. This also led to an honest assessment by management that our job was to create value and the investors' job was to realize it. Having one's share price whacked by events unrelated to our core business model raised the question of just why we were working so hard when the market was giving us no credit.

So, we were casually thinking about going private when these offers came in. We were unable to land any of the big eight deals, and the offers made to us by both Franco and Royal were compelling. Having two or more bidders created perfect deal tension.

When selling one's company, there comes a point where you tell the bidders to sharpen their pencils one last time and make their best and final offer. IRC did this three times! But in the end, Royal Gold made the best offer and IRC accepted. Jensen and the Royal Gold team did a yeoman's job of closing the deal exactly on the fifth anniversary day of our IPO. Gordon Bogden, one of IRC's directors at that time, headed the special committee of the board, which managed the transaction that was successfully completed on February 23, 2010.

In conclusion, IRC was sold for Can$745 million, having begun its life financed with credit cards and having raised less than half of that amount during its lifetime. The company was formed to purchase any royalty, regardless of its underlying metal.

We learned over time that investors will pay a premium for precious metals royalties versus base metal or other commodity types. We also saw the industry pick off most of the major royalties and begin the transformation from royalties to metal streams. Our original quest was to build a billion-dollar company, so we admittedly came short. But along the way, we had a wonderful time gathering up royalties all over the world, building true shareholder value, and being on the leading edge of what is now one of the largest sources for financing new mine developments worldwide. It was a good run.

7

SANDSTORM GOLD

TSX: SSL   NYSE: SAND

sandstormgold.com

Nolan Watson

The story of Sandstorm Gold, how it came to be, what it is, and where it is going is a story of people and a story of values. One evening in the winter of 2008, I was sitting in my car with my young daughter sleeping in the back seat and my newly pregnant wife crying in the passenger seat. Up to this point, she had been stoically understanding of the insane pace of my school and work life, while I was achieving many career and personal milestones. In the first five years of our marriage, I had graduated university (with honors), started my career articling to be a certified public account (CPA) and completed the CPA program (finishing first out of 1,000 people in western Canada), obtained my certified financial analyst (CFA) designation, became the first employee at the world's first streaming company (Silver Wheaton Corp.), was promoted to the role of chief financial officer (CFO) becoming the youngest CFO of any multibillion-dollar New York Stock Exchange–listed company in the world, and founded a humanitarian charity educating children in Africa (an organization I still run to this day).

To accomplish all this, I worked around the clock with little time to sleep and virtually no time to be a good father or husband. The CFO role at Silver Wheaton was my dream job, and I loved every minute of it, especially the people, including Randy Smallwood—who is an amazing human being and leader (now CEO of Wheaton Precious Metals Corp.)—and David Awram, a geologist and one of my favorite people on earth. However, the hundred hour or more workweeks were taking their toll on me and my family. I was raised by value-oriented parents who had taught me the importance of family, and I knew then and there, sitting in that car with a crying wife begging me to find the time to be a proper father, that the status quo wasn't an option. I agreed with my wife and knew that I needed to move on from my dream job, but I had no idea what to do next.

Many people credit me for having the vision to boldly step away from Silver Wheaton and the start-up Sandstorm; but the truth is, that wasn't my plan. In fact, I had no plan other than to be a better father. When I communicated this to the team, it was my coworker and friend, Awram who told me of his plan—he wanted to start up a gold streaming and royalty company and he asked me to join. The idea excited me, and there was no other person I wanted to work with more. So later that year, in 2008, we began our new journey and started Sandstorm.

During our time at Silver Wheaton, Awram and I witnessed firsthand the need mining companies had for financing solutions that were more flexible than traditional methods of debt and less dilutive than equity. We felt we could replicate Silver Wheaton's success while continuing to shape the streaming industry by offering new and innovative solutions to mining companies. Awram and I were optimistic based on what we saw as a fundamental need in the market. Mining companies need financing solutions tailored to their unique capital requirements and long-lead operational hurdles. This is the fundamental reason why the streaming and royalty industry exists.

A mining company's capital requirements are intense, and the enormous amount of capital required is daunting. Given the extraordinarily low number of mines that are built on time or on budget, as well as the tendency for mines to produce less than planned at higher costs per unit than planned, financing the construction of a mine with debt is a terrible idea. Banks are slow, risk averse, nervous, and have the tendency to pull the rug out from mining companies at the worst possible time. The mining investor landscape is full of scarred investors and entrepreneurs who lost everything because they borrowed money from nervous bankers who called their loans prior to the mine achieving profitability. It is a common occurrence in the industry for a mine to achieve profitability a few years after what was originally planned— not only because of cost and time overruns during construction, but also as a result of the nature of complicated plants, which involve hundreds of steps in the process of converting rock into metal or concentrate. If any one of those steps is not working, the mine cannot make money; and because there are so many steps, it is a virtual certainty that not every step will be constructed and operate correctly on the very first try. When a mine first starts up, a long process of working out the kinks begins. This is a normal and healthy part of the process; however, debt bankers don't think this way. Their goal is to simply not get fired from their job. They will happily call a loan if it means job preservation, but this is catastrophic for the mining company and its investors.

> *We are uniquely positioned in the mining industry to touch and improve the lives of people and communities in underdeveloped countries.*

On the contrary, streaming and royalty finance is incentive aligned and lengthy. Our deals are structured to make money over the long haul, and the more long-term success the mine has, the more money the streaming company makes. The industry needs this financing mechanism. We knew it and were excited to start the new adventure that would become Sandstorm.

Unfortunately, our timing could not have been worse. Immediately after beginning our new venture, the financial world decided to have a generational collapse and the Great Recession began. Having no assets and little money, we were entirely reliant on the equity markets to raise the money for us to buy our first streams and royalties. With the world in an epic state of financial disarray, the banks stopped lending, so debt wasn't an option, and equity markets shut entirely with no hope in sight. What the hell had I just gotten myself into?

I now had a daughter, another one on the way, and had just committed to the education of orphans and underprivileged children in Sierra Leone, Africa, with the humanitarian charity I recently founded. Now I had no job, no money because I had put everything I had into the creation of Sandstorm, and what seemed like no hope of being able to raise the equity required to buy our first streams and royalties to start Sandstorm.

Despite all those obstacles, they did not deter us from trying to find mining companies that would want stream financing to help build their mines. At the time, it was still commonplace for junior mining companies to use bank debt to help build their mine, a practice that is less common today, and one I believe will become even less common in the future. It just so happened there were two Vancouver mining companies, Silvercrest Mines and Luna Gold Corporation, which were trying to raise financing to build their first mines, the Santa Elena silver mine in Mexico and the Aurizona gold mine in Brazil, respectively. In conversations with each company, they explained that they were using an Australian bank, Macquarie Group Limited, and both companies told us the same story. They felt they were on a never-ending wheel of bureaucracy and diligence with no end in sight and were willing to consider other options including a stream from Sandstorm. After a month of negotiations, both companies simultaneously agreed to walk away from Macquarie and let Sandstorm provide them with construction finance in the form of a stream, subject to Sandstorm actually being able to raise the equity capital (Can$48 million) to fund the two transactions.

Now, the hard part. How were we going to raise Can$48 million in the middle of the Great Recession with the equity markets fully closed? Despite the dire situation, we found one investment bank, Paradigm Capital, which was willing to help us launch Sandstorm and raise the equity.

On March 12, 2009, we announced our first two gold streams with SilverCrest and Luna Gold and kicked off our initial equity financing. Paradigm Capital accompanied me and Awram on trip after trip, traveling to many different financial cities around the world—to this day, I am thankful to Paradigm for helping us launch. At one point, Paradigm had completed an equity book full of institutional investors and sent out an email informing everyone whom Awram and I had met that the official book in which investors were participating would close the next day. I went to bed that night ecstatic knowing that we had not only managed to land our first acquisitions, but tomorrow, we would become the first company to complete a major financing and launch in Canada during the Great Recession. Little did I know that while I was sleeping, disaster would strike. That night, the gold price unexpectedly dropped $100 per ounce, one of the single largest overnight drops in the history of gold, and our institutional investor book entirely collapsed down to a measly Can$12 million. The financing was dead. Sandstorm was dead. I was done.

"Nolan, you probably don't want to hear this message, but one of the key ingredients to success, is luck." Ian Telfer, the CEO of Goldcorp, once told me that—and he was right. I didn't want to hear it and I didn't believe it, but fortuitously, I eventually found it to be true. I didn't know it at that moment, but luck would be on our side. When we started Sandstorm, the initial shell company came from a venture capital firm called Pathway Capital Management, which had graciously allowed us to use their office space while we attempted to launch the company. At the time, I barely knew the principals, David DeWitt and Marcel DeGroot. What I didn't know then, but am eternally grateful for now, is the enduring character, values, and persistence that they bring to each and every deal they are involved with. DeWitt and DeGroot effectively put Sandstorm's equity financing on their backs and carried it across the finish line, not only writing very large personal checks (much larger than they were originally hoping to put at risk) but also pulling in their vast network of high-net-worth contacts and friends. Pathway Capital dragged our financing across the line, and Sandstorm was born. I will be forever indebted to them, and I can honestly say to any entrepreneur looking to start up a new company in the resource space, you should phone Pathway Capital.

As the Great Recession subsided, we continued to grow Sandstorm by making additional gold stream acquisitions funded by equity raises. I was hell-bent on making every single dollar count. This money was shareholders' money and

I took my role as steward of that money seriously. I put an extreme focus on keeping general and administrative expenses low, spending as little as feasible on full-time employees and using only consultants wherever possible—a strategy that I would soon come to regret. You see, consultants don't have skin in the game, and sometimes their advice isn't worth the paper it is written on. It is a lesson I learned the hard way when we purchased a precious metals stream on a mine in Brazil that would end up at 80% complete in the construction process, only to find out that the dewatering infrastructure that had been built was entirely insufficient. The technical work had not properly planned for the dewatering of the mine to allow full mining operations. Our consultants didn't catch that, and it cost our shareholders, including myself, a lot of money. Outsourcing due diligence was a mistake that I would never make again, and I became determined to build one of the strongest in-house technical teams in the world. It wasn't cheap, and it still isn't, but that team is worth its weight in gold. Sandstorm would go on to make a profit on every single material investment it ever made for the next decade straight (and counting). Technical strength is now a hallmark of the Sandstorm brand. Our team isn't perfect, but they are as close as you will find in this industry. At Sandstorm, we have an entrepreneurial mindset and we believe in failing forward; in this case, we did exactly that.

With success comes imitation and competition.

With a clear vision and a strong technical team, Sandstorm went on to become a billion-dollar-plus New York Stock Exchange–listed company. Despite the dramatic downturn in gold markets from 2012 to 2018, we achieved more than 400% returns for our original shareholders, and yet, we still feel like we're just getting started.

At the beginning of this chapter, I mentioned that the story about Sandstorm is a story about people and a story about values. One of those values that has quietly and persistently pervaded our ethos, is the desire to make the world a better place. We believe in gold and the value that it has to society and monetary systems as well as technology and space exploration, but we also believe that we are uniquely positioned in the mining industry to touch and improve the lives of people and communities in underdeveloped countries.

For far too long, our industry has been one in which foreign companies enter local communities to extract resources at a profit and leave nothing behind but tax dollars, which because of bureaucracy, inefficiencies, and corruption do not make it back to the local communities within which they operate. We know worldwide capital is beginning to flow out of traditional funds and into funds that focus on environmental and social causes and environmental, social, and governance (ESG) criteria has become the buzzword of the decade (and rightly so).

At Sandstorm, these values are deeply ingrained in our people. In fact, since the very beginning of Sandstorm, it is one of the key criteria I have used to hire people. I don't just want smart employees; I want good human beings who care about the world and want to make a difference.

As a streaming company, sometimes that is hard to do because we're not on the ground. We are the financiers, but we are not involved in running the company. However, one thing we do ensure is that when we finance and empower a company to build and grow, they also have the right ethos on the ground too. This desire to make a difference in the world has been within us from the start. It has been in me from the start. It is no mistake that I set up my first humanitarian organization educating children in Sierra Leone before I started up my first business venture.

Quietly behind the scenes, we (me, Awram, Pathway Capital, and Sandstorm's employees) are proud to have built and continue to fully operate, with our own personal money, the Pathway Academy in the city of Waterloo, Sierra Leone, which now has 600 full-time students. Our first graduating class was in 2022 and our students are the pride of Sierra Leone. In West Africa, graduating students write one standardized test for all of West Africa called the West African Senior School Certificate Examination (WASSCE), which determines if children fail (a common occurrence) or pass (most students) or pass with full university eligibility (a high achievement). Pathway Academy students, who are entirely funded by Sandstorm employees and directors, had 100% of their graduating class achieve full university eligibility, a feat that I would bet few, if any, other schools in West Africa have achieved.

At Sandstorm, we are focused on making investors money and achieving better risk-adjusted returns than they could receive if they invested elsewhere, but while we are doing that, we are trying to actually make the world a better place. It is in the fiber of our being; it is who we are. When people invest in Sandstorm, they are investing in a company that will work hard to make them money, but they are also investing in a company that wants to make them money in a way that makes the world better. Capital is finally flowing around the world with an ESG conscience, and Sandstorm and its employees are excited to be a part of that movement. To make a difference in the world, you must have capital willing to follow. Capital tends to follow returns and specifically risk-adjusted returns; therefore our goal is to have the best risk-adjusted returns in the world.

WHERE NEXT?

Over the years, being part of the streaming and royalty sector and watching it grow from virtually nothing to the more than $70 billion industry it is today has been a lot of fun, and it has made investors a lot of money. However, with success comes imitation and competition, and we have seen the landscape of royalty companies expand dramatically. The industry is more competitive than ever, and the rates of return that streaming and royalty companies have been willing to accept have come down across the industry. At the same time, the absolute value of stream and royalty financings completed by mining companies has remained relatively consistent over the last number of years at an average of $2 to $3 billion per annum. With a relatively stable rate of financings and an explosion of new entrants, competition in the industry has intensified. As such, we're seeing deals being made that don't come anywhere close to something Sandstorm would consider accretive, many struggling to realize double-digit internal rates of return (IRRs) and even some deals inked at negative IRRs.

If Sandstorm is committed to earning an above average risk-adjusted return for our shareholders, it begs the questions, "Where do we go from here?" and "Where does the industry go from here?" The answer requires strategic thinking within the streaming and royalty industry. For the industry to grow into the hundreds of billions in collective market capitalization, I believe royalty companies need to think more creatively about the way they acquire streams and royalties. The industry feels a little bit like Netflix's business back in the day when all Netflix did was display other creators' content, which was effectively a commodity business that anyone could replicate. We deploy capital, which any financial institution that has access to long-dated capital can do. Fortunately for Netflix, there was a way out of their problem in vertically integrating and creating their own proprietary content. Unfortunately for royalty companies like Sandstorm, creating our own content is not an option for us. Vertical integration means becoming a mining company and losing all the low-risk positive attributes of being a royalty company.

The best is yet to come for our royalty industry.

At Sandstorm, many investors believed I made this mistake once, by buying a minority equity interest in a project (Hod Maden in Turkey) despite it being a world-class asset. I am rectifying this by restructuring it into a stream, and Sandstorm will not ever directly vertically integrate by becoming a mining company. However, there is an advantage royalty companies have that has not been sufficiently utilized. Our corporate development and technical teams are very large, and we see an enormous amount of deal flow. Anytime a mining company is thinking about selling or financing an asset, royalty companies get a call. It isn't very often when someone calls us regarding a mine that we haven't already seen or done due diligence on over the years, and we are virtually always in the industry's deal flow. We see almost everything that is happening in the industry, and Sandstorm now has deep pockets to finance larger mine acquisitions. Because of this, I envision a future where every mining entrepreneur and growth-oriented CEO will have a royalty company joined at the hip for the first several years of their growth cycle until they are large enough and diversified enough to raise their own equity debt and equity capital, no longer requiring their royalty partner.

When I first got into the mining industry, equity capital was abundant. Everywhere you looked there were mining-specialty equity funds ready and willing to write checks into bought equity deals. Now the world has changed. Capital has been removed from these mining-specialty funds and most have been shut down or are a shadow of their former selves, with the capital being put into large, passively managed ETFs instead. These ETFs do not write equity checks into financings, which means the capital is no longer available to the next generation of mining entrepreneurs who want to start and grow their businesses. The need for royalty finance is greater than ever, and it is a structural change in the capital markets that I expect to persist indefinitely.

At Sandstorm, we're not just looking for good deal opportunities, but also for entrepreneurial-minded CEOs who are seeking to grow their companies. I envision a future where every growth mining company will have a royalty and streaming partner to assist them in both finding the mines that are for sale and helping finance the purchase. Unlike traditional banks, our industry has the expertise and relationships to ensure that smart deals are made, offering better outcomes for mining companies, and better returns for investors. I truly believe the best is yet to come for our royalty industry.

8

OSISKO GOLD ROYALTIES

TSX | NYSE: OR

osiskogr.com

Sean Roosen and Sandeep Singh

OSISKO GOLD'S FORMATION

Osisko Gold Royalties was formed as part of the sale of the original Osisko Mining Corporation, starting with a 5% net smelter return (NSR) royalty on the world-class Canadian Malartic gold mine in Québec, Canada. Osisko Mining Corporation was formed by Bob Wares, John Burzynski, and Sean Roosen in 2003. Roosen and Burzynski spent the 1990s working in Africa, and Wares had been working uniquely in Québec as a consultant and running a Toronto Stock Exchange Venture Exchange (TSXV) exploration company. The three partnered in Montreal to hunt for low-grade bulk tonnage gold mines in old Canadian mining camps, principally in areas that were considered mined out, and focused on assets that had the potential to be world class in size. Wares thought that there should be gold-only porphyry deposits in the Archean gold belts, and the trio successfully proved the point with their delineation and subsequent production of the Canadian Malartic gold mine and their discovery of the Odyssey deposit, with more than 30 million ounces of gold combined. They are one of the few teams in recent Canadian mining history to take a company from formation, through concept, discovery, financing, development, and into successful and profitable production. Twenty years after they started, Canadian Malartic is the ninth largest gold producer globally and remains Canada's largest gold producer.

The story began with the initial purchase of the old Canadian Malartic property from a bankruptcy sale for $80,000. The group then conducted a large-scale drill program to define what has now become the Canadian Malartic open pit mine. From 2008 through 2009, Osisko created the new modern neighborhood in the town of Malartic by relocating 300 homes, and built six new institutional buildings, which were handed over to the government and community. These included the most technologically advanced primary school in northern Québec, a community center, a supported living residence, and a recreation center. Following successful capital raises in excess of Can$1 billion in 2009, the Canadian Malartic mine was permitted, built, and put into production in April 2011. Osisko then set out to optimize the mine, but was interrupted in January 2014, when Goldcorp launched a hostile bid on the company, and the direction of the company was permanently altered.

The main goal of Osisko Gold Royalties was to increase precious metals exposure for its shareholders through the addition of high-quality royalties and streams.

At the time, Goldcorp was the largest global gold company, with the best balance sheet and the highest trading multiples. The bid, however, was opportunistic and heavily undervalued the company. After a lengthy battle, a friendly transaction was announced with Agnico Eagle Mines Limited and Yamana Gold where they each acquired half of the mine.

As part of the transaction, management knew that they were leaving value from the mine's underground potential behind as initial underground drilling had discovered mineralization at depth. To retain exposure to Canadian Malartic, and in particular, the potential for a new world-class underground extension, a 5% NSR royalty was spun out along with Can$157 million in cash to create Osisko Gold Royalties. The retention of a meaningful royalty allowed Osisko shareholders to retain exposure to Canadian Malartic and its prospective upside. Since the completion of that transaction in June 2014, Osisko Gold Royalties has more than quadrupled its market value while amassing a portfolio of more than 165 streams and royalties, and the resource base of its flagship asset, Canadian Malartic, is larger today than it was when sold in 2014.

SPIN OFF VERSUS SELLING OUT

In the years leading up to 2014, management of Osisko had been singularly focused on bringing Canadian Malartic into production and ramping up what has become Canada's largest gold mine. Prior to Osisko, the prospect of hard-rock gold mining of perceived low grades in Canada was not well accepted. In fact, there were some that used to refer to the Canadian Malartic development project as Bre-X North!

Similarly, when the supported sale was announced with Agnico and Yamana, there were some who wondered whether the underground drill results were merely an attempt to extract incremental value. In reality, the underground potential was always something that the team had intended on investigating. Having run out of runway because of the hostile takeover, it was deemed smarter to retain as much exposure to the open pit mine, and in particular, the underground potential, as possible.

The resource base of Osisko's flagship asset, Canadian Malartic, is larger today than it was when sold in 2014.

As the hostile process unfolded with various interested parties, management kept working on plans for a SpinCo royalty company and ultimately were successful in negotiating that SpinCo with Agnico and Yamana. The significant competitive tension created in the process, in addition to the buyer's focus on the mine itself allowed Osisko to extract significantly more value for their shareholders.

The SpinCo, which is today Osisko Gold Royalties, started out with a 5% NSR on the Canadian Malartic open pit and four other exploration-stage royalties in Ontario. Since then, management has been proven correct with the Odyssey underground project at Canadian Malartic and now comprises approximately 15 million additional ounces and is still growing. Had it not been for the SpinCo, that value would have been lost to Osisko share-holders, or forfeited for very little, if any, up-front value. The additional exploration royalties too are shaping up nicely in the Kirkland Lake camp (Ontario, Canada).

> *Osisko chased a short list of ideas and formed new public companies in the process, each time seeding the new ventures with technical and business talent from its founding group.*

In addition to the royalty assets, the SpinCo retained Osisko's proven tech-nical team—the team that discovered, permitted, built, and operated the Canadian Malartic mine. Today, that technical team continues to advance the royalty business as well as having ventured out to start affiliated entities with some of the better development and exploration stories in the sector, including the world-class Windfall deposit in Osisko Mining, among oth-ers. Like Canadian Malartic, Windfall has quickly become one of the largest deposits ever defined in Québec (Canada). It stands as the largest high-grade underground deposit discovered in the province to date.

ORION PORTFOLIO PURCHASE

From the onset, the main goal of Osisko Gold Royalties was to increase precious metals exposure for its shareholders through the addition of high-quality royalties and streams. Osisko needed to diversify its asset base and cash flow to cement the company as a real royalty player. When the Orion Resource Partners mine finance portfolio became available, it made a good fit. The transaction would be transformational for Osisko. The acqui-sition added at the time 74 assets, including 61 royalties, 7 precious metal offtakes, and 6 streams, diversified the company's existing cash flowing asset base, and enhanced the company's development and exploration pipeline of assets. Pro forma, Osisko held more than 130 royalties and streams, including 16 revenue-generating assets while remaining largely Americas focused.

The acquisition also allowed Osisko to strengthen its ties to the Caisse de dépôt et placement du Québec and the Fonds de solidarité FTQ who together subscribed for Can$275 million in common shares of Osisko Gold and increased Osisko's market presence and trading liquidity.

The acquisition essentially doubled Osisko's market value at the time and provided the company with the increased scale and diversification needed to compete in a growing royalty space. The deal was widely lauded by the investment community with Osisko winning Large-Cap Mining Deal of the Year at the 2017 Mines and Money international conference.

PROJECT GENERATOR MODEL

The concept behind the accelerator model stems from the extensive amount of technical and capital markets expertise that Osisko Mining carried. The Osisko team had the opportunity to review and conduct due diligence on a great number of assets. No longer running the Canadian Malartic mine, they had the bandwidth and a head start on finding Canada's next generation of mines, Windfall being a prime example. They set out to chase that short list of ideas and form new public companies in the process, each time seeding the new ventures with technical and business talent from its founding group. To start these ventures, Osisko Gold Royalties also provided the seed capital in return for royalties and streams on attractive exploration assets.

This symbiotic relationship allowed the group to form about one new company a year since 2014 and has led to a number of the most interesting discoveries in the mining space. In the process, the Osisko group has created value for Osisko Gold Royalties shareholders and the stakeholders of these new public vehicles and helped secure an unparalleled growth pipeline for Osisko Gold Royalties without having to pay the most in a field of larger incumbents.

9

METALLA ROYALTY & STREAMING

NYSE | TSX: MTA

metallaroyalty.com

Brett Heath

SETTING THE STAGE

It was early 2016. We didn't know it yet, but a brutal four-year bear market in gold and gold equities was over. The bottom was technically $1,060 per ounce in December 2015. After running from approximately $300 per ounce to more than $1,900 per ounce the decade prior, gold took a few years to find its footing. However, when it finally turned the corner, it ran from $1,060 to nearly $1,400 per ounce in a hurry.

By the summer of 2016, the mining sector, known never to waste a bull market, was red hot. It seemed like every company with a pulse raced to issue stock. That's how things go in the gold business. Capital flows run like a spring flood in both directions. It was the first time investors wanted anything to do with the gold business in four years.

The major royalty companies took advantage of the four-year bear market. They emerged in a different state. For instance, it was the first time we saw a streaming transaction top $1 billion.

Major mining firms didn't have the same experience generally. Many came out of the bear market with serious trouble on the balance sheet. During the decade-long bull market of the 2000s, they stretched, paying whatever it took to buy growth.

Another critical piece of the puzzle made this upswing different from any before. It set the stage for the royalty and streaming titans who'd dominate the coming decade. You see, for the good part of a century, senior-lending banks financed mine building. However, they made mistakes. The downturn left them tending to big losses on risky loans. That left a gap royalty, and streaming firms stood by, ready to fill.

The royalty and streaming sector invested almost the same amount of capital as the entire Toronto Stock Exchange Venture Exchange (TSXV) raised during those same years (2013–2015). It was incredible to watch.

I worked in the royalty business from 2012 to 2016 as a consultant. I had great relationships in the sector from my years as a fund manager. Even with a strong network, fundraising was impossible without a meaningful track record and loyal investor base. This was not a time for upstarts. That said, we had a great idea. It was a big idea. In retrospect, it changed the lives of some early investors. It changed the industry and generated a half dozen copycats in a matter of years.

ALCHEMY IN THE ROYALTY BUSINESS

We knew the royalty business was the only way to play gold's next great bull market. The only problem was that we didn't have any royalties. We also didn't have much capital. To overcome these two seemingly critical missing pieces, we came up with the idea that took us from zero to more than $500 million in half a decade.

We felt confident that the trend change we saw in early 2016 was real. This was the right time to start a new royalty company. With the gold price stabilized, we needed to buy enough royalties to give us a foundational portfolio before confidence returned to the market.

As a consultant, I structured many royalty transactions with junior producers. While some succeeded, many did not. Some had bumpy landings. The value of the royalties suffered. In my view, the royalty owners, urgent to get deals done, often sacrificed the key features I felt made royalties so valuable. I told my backers and my early team we didn't have that luxury.

We'd need a different approach to be successful. We'd need our approach to be scalable. Remember, this was a big idea, not a small one. Supervising fledgling miners would bog us down. It wasn't an option.

Then we had a money problem. Some firms had huge checkbooks to finance growth. We had a small one. We needed that small checkbook to buy us exposure to major projects. This means top-tier operators working the proven geological trends. The apparent problem was that no major mining company would ever encumber their assets for a small dollar amount or any amount for that matter.

Here's where our plan carved out a brand-new track. We'd go directly to third-party owners of premium royalties. We'd offer them stock in our new firm in exchange for their valuable royalty. They'd have an increasingly liquid stake in a diversified holder of royalties.

We'd also explain that this diversified collection could value their royalty at a premium to their expectations. After all, they had one royalty; here, they'd have some insulation from operating issues, jurisdictional changes, and the occasional streak of bad luck that can hit any miner.

My investors thought this was too good to be true. I showed them there was a secondary market of royalties. I'd been making a list of these orphaned royalties for years. Of the thousands out there with non-royalty company owners, I thought I'd be able to target at least 100. When I showed them the list, most of my investors were all in.

To be clear, the universe of orphaned royalties exists for a very logical reason. Most of these came into existence when a prospector found something years ago. Nothing happened for a long time. When it did, a mining firm usually bought the asset, leaving the prospector with a royalty. If the acquiring firm was a major mining company, we wanted to talk.

We realized this market was much bigger than we appreciated as we got started. Yet we were the first royalty company to make this market a sole focus.

Almost every gold property on the planet had a royalty. These projects often go through various owners from discovery to production. Most of these owners retain a royalty as the project advances from exploration stage to commercial production. Established producers capable of building and operating the mine would rarely agree to initiate a royalty. We'd target the prior owners who kept a royalty when they sold the project.

We'd have to make it appealing for these royalty holders to let go of their prized asset. We'd need to offer something better than a big check. A platform offering daily liquidity, public market premiums, and diversification might convince them to part with these rare assets. It certainly would be a unique pitch.

If it worked, we'd build a rich portfolio of high-quality royalties by exchanging equity in our new company for these amazing assets.

DEFINING THE PLAN

Many of the best-known royalty companies formed as spinouts from major mining companies. In this case, the mining firm saw these non-core assets as more valuable in a stand-alone entity. Shareholders of the mining firm could often automatically become shareholders of the new spinout royalty firm.

The spinout gives the new royalty company a head start. It provides a foundational asset base, diversification, cash flow, and a stable balance sheet. These are essential ingredients for success in the royalty and streaming business. While we had a big idea, we didn't have any of these building blocks.

Starting out at a disadvantage, we'd have little room for error. We spent months building a pipeline of the best royalties in the secondary market. What started in the thousands narrowed to around 100 coveted royalties. We felt this premium list had the most upside. We started building long-term relationships with these holders. Some of those initial sellers became shareholder partners. Many still provide us with a material advantage today, five years later.

Most of the royalty holders we contacted thought our idea was a great. Few wanted to actually take a risk. Letting go meant trading their coveted asset for shares of a Canadian Security Exchange (CSE)-listed upstart company. The typical response was, "This is a great idea, and I am interested, but I would like to see you go out and get a few deals done before we transact." We didn't take offense.

After a few months, we realized we needed to go public. Liquidity was a key objection to our pitch. This was especially the case with owners of long-dated development royalties.

There were a few ways to go public in Canada. An IPO is the most well-known way. Because our big idea was offering shares to selling royalty holders, raising cash through a public offering didn't help us. A similar approach is a direct listing. This means existing shareholders one day see shares trading under a ticker, with no company shares sold on day one. Neither of these traditional methods solved our problem. We opted for a reverse takeover (RTO).

The RTO is a common route for Canadian companies to tap the public markets. Essentially, the new firm finds a failed penny stock that hasn't yet been delisted. Out of money and options, the controlling shareholders are open to the idea of hanging on to a small portion of the reconstituted firm.

Here's how it worked for us. We found an exploration company called Excalibur Resources. It had a market cap of around Can$500,000 because it often went days without a single trade; it had few options at restarting its business.

I offered the company's board an RTO proposal in July 2016. Excalibur would acquire my private consultant company with shares and give up control of the board and management. In conjunction, we consolidated the old shares, and for every three shares held in Excalibur, they'd receive one share in the new company. We closed the deal on September 1, 2016.

As news of the deal started flowing, investors bought up the old Excalibur shares. Between the agreement in July and closing in September, the new royalty and streaming company with little more than a big idea had a valuation nearing Can$10 million.

TIME TO GET TO WORK

When the deal closed, I stepped in as president with the right to nominate a new board and rename the company. We quickly changed the name to Metalla Royalty & Streaming. The board part took a bit longer.

With the share consolidation complete, we had around 33 million shares outstanding and a clean slate. I brought in my longtime friend and previous business partner E.B. Tucker as an independent director. At the same time, I added industry veteran Lawrence Roulston. Both officially joined the board in early March 2017.

At the time, Tucker worked for Doug Casey as editor of *The Casey Report*. He spent prior years writing other major financial newsletters for Stansberry Research, Bill Bonner, and other major publishers. He brought a unique skillset to the board. In addition to years as a published financial analyst, he knew how to tell our story to new shareholders. Today, this continues to give the company an edge over its peers.

People had a lot of questions about the name. *Metalla* is Latin for precious metal, and Latin is the root of our language. I thought it was a fitting name. New shareholders confused it with the heavy metal band Metallica, which turned out to be an advantage in the end. It was easy to remember.

We cut our first deal in the fall of 2016. A veteran prospector well known in Timmins, Ontario as being a challenging individual to transact with had a portfolio of royalties. One of them covered an extension of the Hoyle Pond mine (Ontario, Canada). This was a major asset for GoldCorp, which Newmont Corporation later acquired.

The Hoyle Pond mine at that time made up about 11% of total tonnage coming out of the Timmins camp. More impressive was its grade—11% of tonnage yielded over 60% of the total ounces coming out of the camp. The main S-vein deposit at Hoyle plunged toward the northeast, drifting directly toward our new royalty property.

This portfolio checked the boxes. A major mining company operated it. It sat on one of the most recognized gold trends in the world, the Porcupine-Destor trend in Ontario, Canada, which hosted many of the world's best gold mines. It had a significant amount of potential growth with no financing risk. The prospector liked us, and he seemed open to taking a chance with our new stock.

A GREAT BUSINESS IN THE SHADOW OF A TERRIBLE INDUSTRY

In creating our big idea, we wanted to figure out exactly what drove some royalty shares far higher than others. We found a few key catalysts rang true across all the industry success stories. Having a major, new discovery on a royalty turned out to have the biggest long-term positive effect on shareholder value.

Great examples of this are the Goldstrike royalty (Nevada, United States) and Detour Lake (Ontario, Canada) royalty for Franco-Nevada Corporation, both acquired for $2 million. The same goes for the Cortez mine (Nevada, United States) royalty, which Royal Gold acquired for $8.5 million. All of these remain some of the most valuable royalties (not including streams) in their respective portfolios after decades of operational success. We found that most of these assets had similar characteristics before they were known. This became a critical piece of how we view royalties and which ones we want to buy.

When average people make an extraordinary profit in life, they call it a gold mine. The wealth created almost overnight when you find an actual gold mine can be astronomical. That's the case regardless of lulls in the gold price.

The problem is that 99.9% of all gold projects become nothing but a hole in the ground. Companies light money on fire while promising the world to speculators. Most of these adventures end in tears.

We needed to increase our odds of business success. Using our strategy, we could cut out almost all the high-probability failures. We could then increase our odds even further by having upward of 100 royalties over time instead of just one.

We also noticed a consistent value curve for pipeline assets within major mining companies' portfolios. We call these development assets. They stand past the point of discovery, and before mine building begins.

Currently, major mining firms rarely use outside financing to fund new mine development. Unlike a smaller, junior development project, they can write checks as they go. The smaller firms need to continually prove what they have to investors along the way. This means being beholden to markets for ongoing equity, debt, or stream financing. Major mining companies, however, complete the necessary technical studies internally before they decide to build the mine.

We asked our investors to picture this discovery-to-production time line as a curve on a graph. As time passes, the curve moves higher. Certainty builds value as time passes. We wanted to capture the most valuable time on this curve, acquiring royalties right before news of major capital spending. To be clear, right before often means two to five years before production. We felt the odds of seeing reserves and resources increase by two to three times was good in that critical period just before construction began.

In our view, the strategy was obvious. Major mining firms plowed tens of millions of dollars into drilling projects just before making a production decision. They knew the gold was there when they bought the project. Surely, sinking hundreds, sometimes thousands, of drill holes in the ground would find something. After all, the best place to find a gold mine is next to a gold mine. Holding a passive royalty claim meant anything found in this expensive treasure hunt boosted the value of our royalty. Our investors agreed.

In the end, if our big idea worked, we'd cobble together a vast portfolio of royalties covering prolific mining projects around the globe. We'd spend very little doing it. If it worked, we'd see the major mining firms of the world advance projects on our claims. They'd invest hundreds of millions of dollars every year. Whatever they found would boost our royalty value. That includes shares held by the people who sold royalties to us in the early days. If one of those miners found something huge, we'd capture that value across our entire share base.

THE EARLY DAYS WERE NOT EASY

As I write, we're six years in. We've consistently executed our strategy. Consequently, we've been the most active royalty company. We completed 29 public acquisitions acquiring 71 royalty and stream assets.

Initially, Metalla had just one employee, me. I worked from home with a newborn. With no existing portfolio of royalties, no corporate financial backer, and no staff, we had an uphill climb. While our competitors had all the trappings of corporate success, we had the big idea.

Marketing the new royalty company proved difficult. We were a true micro-cap stock. To make things worse, we traded on the CSE, which was at the time a down-market exchange. Almost every investor we met with did not take us seriously. We heard every excuse. The typical responses were "you are too late," "you will have too high of a cost of capital," and "you will never be able to compete with the big royalty companies."

We completed our first financing at a share price of Can$1.20 (adjusted for share consolidation) in November 2016. My good friend Adrian Day funded more than half of the Can$2.3 million raised. Adrian sub-advises the Euro Pacific gold fund in addition to running his own asset management business. Casey, an industry icon, and Tucker, my soon-to-be board member stepped in with important support. The balance came from friends and family.

After acquiring the Hoyle Pond extension royalty, we picked up a few more royalties until the portfolio acquisition with Coeur Mining. The Coeur portfolio was a big step in the growth for Metalla because it gave us a meaning-ful cash-flowing royalty with the Toho Endeavor mine (New South Wales, Australia) and a few good royalties on development assets. We acquired the portfolio from Coeur for $13 million.

We structured the Coeur transaction as a share/convertible debenture transaction. This allowed us to complete the acquisition keeping Coeur at 19.9% ownership. That meant it would maintain the 19.9% interest automatically through the convertible debenture as we grew the company.

Mitch Krebs was the CEO of Coeur Mining at the time. I met him about a year before at the Roth Capital Partners conference in Dana Point, California. He was one of the very few that understood our big idea. He believed in our ability to execute the business plan. He understood how his company stood to profit if we succeeded. His bet paid off. After about three years, Coeur successfully sold its entire investment, realizing over $40 million in proceeds.

THE TRACK RECORD

After completing the Coeur transaction, we had some cash flow. We used that to build out our team, board, and go after more targets in our pipeline. We added Drew Clark as vice president (VP) of corporate development and Sunny Sara as an associate analyst. Sunny later became VP of acquisitions. Clark, Sunny, and I continue to make up the deal team today.

Denis Silva, a young associate from the international law firm Gowling WLG became our external general counsel. We worked together on hundreds of different deal structures as each royalty we bought was unique. Having a great legal team is a key part of what made Metalla the most productive royalty company by number of transactions. Today, Silva remains a key part of the business and is now partner of DLA Piper global law firm.

We strengthened the board adding Alex Molyneux and James Beeby with industry-specific company building and legal expertise, respectively. We added technical advisers Frank Hanagarne and Charles Beaudry to quickly help vet opportunities in our growing pipeline. Finally, we added royalty industry veteran and mining hall of famer Douglas Silver who was well known for selling his company, International Royalty Corporation, to Royal Gold for more than Can$700 million and the over $1 billion sale of the royalty portfolio he assembled within Orion Resource Partners mine finance to Osisko Gold Royalties.

As the company and the track record grew, we got the attention of Beedie Capital, which is a family office controlled by real-estate investor Ryan Beedie, who is one of the largest holders of industrial real estate in Canada and has a great reputation for being a very long-term supportive shareholder. Although Beedie Capital backed very few management teams in the mining sector, the similarities between owning royalties and real estate made it easy for Beedie to understand. We ended up striking a deal for a Can$13 million convertible loan facility, which we later expanded to Can$32 million. Today, Ryan Beedie, through Beedie Capital, remains one of our largest and supportive shareholders and was key in funding a significant portion of growth.

After the first five transactions, people started to take us a bit more seriously. They saw we had an effective and efficient process. Our initial track record was well above average. Over the following years, we streamlined our process to acquire existing royalties and quickly scaled up the business with 29 transactions shown in Table 1 from newest to oldest.

Almost all of our deals originated on a bilateral basis. This means we rarely bought assets through traditional channels such as investment bank sales processes. We built our pipeline on the phone, in person, and sometimes by asking a royalty owner to please just consider our big idea before making another decision. Nobody else consistently did it this way.

This set us apart from peers. It allowed us to bring in great assets at accretive prices. In fact, after five-and-a-half years and 71 royalties acquired, we are the only company that has never had to take an impairment charge. That means we've never stretched to get a deal done only to have auditors later devalue the asset on our books. In the hyper-competitive royalty industry, that's rare. Metalla had a clear first-mover advantage and was able to pick up the best assets in the secondary market, specifically in the sub-$20 million price range.

Our key competitor was Maverix Metals, a new royalty company founded by a good friend, Dan O'Flaherty, a few months before Metalla was officially started. A portfolio spinout created Maverix out of Pan American Silver's royalty portfolio. Maverix was the most successful smaller firm in targeting non-core royalty portfolios held by mining giants including Goldfields,

Table 1 Public acquisitions by Metalla Royalty & Streaming

Date	Mine	Company
October 5, 2022	Lac Pelletier (Québec, Canada)	Maritime Resources Corporation
February 2, 2022	Beaufor (Québec, Canada)	Monarch Mining Corporation
September 9, 2021	Castle Mountain (California, USA)	Equinox Gold
June 17, 2021	Côté (Ontario, Canada)	IAMGOLD Corporation
May 15, 2021	Fortuna 2.0 (Mexico)	Minera Alamos
March 20, 2021	Del Carmen (Argentina)	Barrick Gold Corporation
March 16, 2021	CentroGold (Brazil)	Oz Minerals
March 15, 2021	Tocantinzinho (Brazil)	G Mining Ventures Corporation
February 16, 2021	AK and North AK (Ontario, Canada)	Agnico Eagle Mines Limited
November 4, 2020	Genesis (Nevada, USA)	Newmont Corporation Nevada Portfolio
September 16, 2020	Higginsville (Western Australia)	Karora Resources
July 27, 2020	Fosterville (Victoria, Australia)	Agnico Eagle Mines Limited
June 22, 2020	Wharf (South Dakota, USA)	Coeur Mining
April 24, 2020	Goldrush South (Nevada, USA)	Nevada Gold Mines
February 18, 2020	NuevaUnión (Chile)	Teck Resources Limited/ Newmont Corporation
August 7, 2019	Fifteen Mile Stream (Nova Scotia, Canada)	St. Barbara Limited
April 1, 2019	Canada/Mexico Portfolio	Alamos Gold
February 4, 2019	Fifteen Mile Stream (Nova Scotia, Canada)	St. Barbara Limited
January 4, 2019	Dufferin (Nova Scotia, Canada)	Aurelius Minerals
December 11, 2018	COSE (Argentina)	Pan American Silver
September 5, 2018	Santa Gertrudis (Mexico)	Agnico Eagle Mines Limited
May 14, 2018	Akasaba West (Québec, Canada)	Agnico Eagle Mines Limited
May 10, 2018	Garrison (Ontario, Canada)	Moneta Gold
June 12, 2017	Diversified portfolio	Coeur Mining
May 4, 2017	Hoyle Pond Extension (Ontario, Canada)	Newmont Corporation
February 28, 2017	Portfolio (Timmins, Ontario, Canada)	2090720 Ontario
January 18, 2017	New Luika (Tanzania)	Shanta Gold Limited
November 2, 2016	Portfolio	International Explorers and Prospectors
October 15, 2016	Mirado (Ontario, Canada)	Orefinders Resources

Newmont, Kinross, Goldcorp, and Pan American Silver. Metalla led the way with the individual holders and managed to acquire a couple of portfolios from Coeur Mining and Alamos Gold.

We used our competitive advantage and developed great relationships with the royalty holders in the secondary market we wanted to turn into shareholders through buying their royalties. Most transactions had significant lead times, often lasting longer than a year. We continued to refine our target list over time, and if we had enough conversations moving forward at once, we found that there was usually one of them that was ready to transact.

This is not a strategy that can be replicated by focusing on the same commodities. Once you buy a royalty in the secondary market, it is removed. Most royalties we bought were created in the 1970s through 1990s. Even with an unlimited budget, we would not be able to replicate the portfolio we built over the last six years.

Because of this, all new entrants that used the strategy of buying royalties in the secondary market between 2018 and 2022 struggled to create shareholder value. They found themselves forced into competitive processes, often paying top dollar for lower-quality royalties. Some settled for royalties on other types of base metals and commodities. Both of those strategies went against everything that makes the gold royalty business what it is.

GROWING UP

While Metalla started on the CSE, we had a goal to graduate to the New York Stock Exchange (NYSE). People laughed at our goal. In fairness, it did seem miles away with a single-digit million market capitalization.

While investors saw our portfolio of royalties expand, internally, we worked hard toward a better listing for our stock. First, we moved to the TSXV. This was a big step up. By February 2018, we made it happen.

The U.S. markets were significantly larger and more liquid. With the help of Tucker, we tapped into various U.S. retail channels. They liked him, and they loved our story. We turned over our existing shareholders left from the company's RTO. Over time, new U.S.-based investors made up a growing portion of our shareholder base.

From Can$10 million early on, our market capitalization reached nearly Can$200 million in early 2019. A key requirement for NYSE listing is $75 million in market capitalization. Shareholders owning more than 10% of your stock are excluded from the calculation. Coeur held 19.9% at the time. That meant we had an extra hurdle to clear.

We brought on Saurabh Handa as chief financial officer and Jonah Townsend as controller to build out our internal finance team. We needed to play a little catch-up given how fast the company was growing. Under Handa's leadership, it took no time at all.

We completed all the requirements and began trading on the NYSE in December 2019. In the prior year, we visited just about every major city, pitching the company to potential shareholders. Most U.S. brokers were not allowed to buy stocks listed in Canada for their clients. This created a significant amount of pent-up demand. Tucker's media presence contributed too. We saw the stock rise from Can$5.38 to Can$7.18 (adjusted for share consolidations) the week of the NYSE announcement. It was a huge success and validation of our big idea.

The NYSE asked us to consolidate our stock in a 4-for-1 split, meaning for every four shares held prior to listing, shareholders would have one. The exchange wants to avoid shares trading below a $5 threshold as some institutions are not allowed to own lower-priced stocks. We were safely over that level and positioned Metalla for continued success.

BUILDING A TOP-TIER PORTFOLIO

When buying royalties out of the secondary market, there is almost always a story behind it. Every transaction is different, with various twists and turns along the way. We bought royalties from corporations, individuals, or families with history with the projects, famous billionaires, geologists, exploration companies, lawyers, crypto companies, and everything in between. We often had to develop unique structures to create a win-win scenario for our acquisitions. We aimed for every new asset to add value for existing shareholders.

Over the first six years, we deployed approximately Can$181 million in the 71 royalties we acquired. Some of the key royalties acquired were:

- Santa Gertrudis (Mexico), Agnico Eagle
- Endeavor (Australia), Toho Zinc
- Joaquin/COSE (Argentina), Pan American Silver
- Wasamac (Canada), Yamana Gold
- Fosterville (Australia), Agnico Eagle
- Goldrush South (United States), Nevada Gold Mines
- Garrison (Canada), Moneta Gold
- Wharf (United States), Coeur Mining
- Del Carmen (Argentina), Barrick Gold
- Castle Mountain (United States), Equinox Gold
- Amalgamated Kirkland (Canada), Agnico Eagle
- Côté Lake/Gosselin (Canada), IAMGOLD
- Tocantinzinho (Brazil), G Mining Ventures Corporation

We have had an excellent track record of creating value for our shareholders. Each new asset brought into the business reflects a boost in value through the public markets.

We would fund new transactions as we were ready to transact. This often provided a great check and balance as investors had to be in support of the deals. We were able to do this at consecutively higher prices (Table 2).

In addition to money raised from the capital markets, we issued approximately Can$100 million of shares directly to the royalty holders based on a volume-weighted average price to closing the transaction. This aligned all the existing and new shareholders of the pro forma company.

Table 2 Funding new transactions

Date	Price, Can$ (adjusted)	Amount, Can$ million	Type
November 30, 2016	1.20	2.37	Private placement (PP)
March 6, 2017	2.00	3.20	PP
April 10, 2017	2.00	3.20	PP
January 4, 2019	3.12	6.78	PP
April 1, 2019	5.56	7.00	Convertible debenture (CD)
June 22, 2020	7.21	28.19	Secondary
August 30, 2020	9.96	5.00	CD
May 14, 2021	12.17	21.59	At the market (ATM) facility 1
May 13, 2022	10.36	20.48	ATM 2

MOVING FORWARD

The success of Metalla, and key competitor Maverix, generated a lot of copy-cats. Around half-a-dozen upstarts flowed into the industry. Many claimed a similar type of strategy.

By 2021, more royalty companies than ever existed in the history of the business. Many struggled to grow given the increasingly high level of competition. We saw valuations in competitive processes over this period rise dramatically. The investors became a bit tired from all the new entrants. In short, the business became over-competitive making it nearly impossible to complete accretive transactions if they were competitive.

By 2021, we saw the tide start to shift. We knew we needed to bulk up and find a potential cornerstone asset. By year's end, 2021 was the best year Metalla ever had on the transaction front. We completed seven acquisitions and brought in royalties that covered 235,000 ounces of gold to our account for about $40 million with two high-profile potential cornerstone assets with royalties on Côté Lake/Gosselin and Castle Mountain. That is $170 per ounce, well below the industry average on very high-quality assets.

Côté was always a big sleepy project to the industry participants. In 2019, they discovered Gosselin. Sunny called me and said, "I have a new deal for us!" He spotted incredible drill intercepts coming from this new area, and by early 2021, we knew there was something special there. Several multi-hundred intercepts of 0.5 to 2 grams per metric ton gold grades were near the surface. We quickly went back to our list and found a royalty held by a few private parties covering the entirety of Gosselin and a small portion of Côté pit reserves.

The holders of the royalties were way off the radar. After some digging, we made the connections and started the process. A few months later, we acquired a 1.35% royalty from two private sellers on the project for Can$7.5 million. We completed this acquisition when there was no defined mineral resource.

The key to buying great royalties is to get them before anyone else understands their potential, including the operator. Once they are known, they are rarely sold. If they do sell, it's for an insane price, which is far more than we can justify to our shareholders. This requires a calculated risk that Metalla would take, a risk that we well understood after buying 71 royalties over the previous years.

We believe Côté/Gosselin combined will become one of the top five gold mines in Canada and serve as a key cornerstone asset in the portfolio over the next few years.

In my view, 2020 kicked off a new commodity bull market. We saw commodities broadly make a significant advance. Many doubled over the following 18 months. We also noticed the major mining companies across the board had less leverage on their balance sheets. They began disciplined and necessary mergers and acquisitions.

This meant far fewer new royalties and streams coming into existence. There seemed to be still plenty of single-asset companies needing capital, but those often come with considerable risks. The lower-quality process-related transactions showed up constantly as the sector was in a severe seller's market.

Some smaller, more vulnerable royalty companies started combining during this time. Gold Royalty Corporation acquired Ely Gold Royalties, Golden Valley Mines and Royalties, and Abitibi Royalties. Nomad Royalty Company and BaseCore Metals were acquired by Sandstorm Gold. Elemental Royalties and Altus Strategies merged. We saw the industry starting to eat itself. More importantly, we saw a large portion of our peer group disappear, which provided a great opportunity for Metalla to continue on its path of growth through the next bull market cycle in gold.

The sentiment around gold and silver, which were the worst-performing commodities in 2021, was at bearish extremes. You couldn't have dreamed up a better macro backdrop for the metal. As a group, the equities saw a slow bleed of investor outflows from the summer of 2020 through May 2022. Metalla and many other equities were significantly off their highs as investors slowly lost confidence.

As the tide shifts again, we believe Metalla is positioned as a coiled spring for the next bull market. The royalty sector is continuing to have the pressure of consolidation, and there's a scarcity of high-quality assets. As this trend picks up steam over the next couple of years, we see Metalla in a significant position of strength: our built-in growth across 71 royalty assets, with a dozen being advanced by majors into production every year for the next decade. We see significant improvement of our emerging cornerstone asset on the Côté/Gosselin mine. We look forward to increasing potential and early signs of a major discovery on one of our royalty properties with so many being actively advanced at the cost of the major operators. Today, Metalla sits on an enviable foundation ready for the next chapter to begin.

10

GOLD ROYALTY CORPORATION

NYSE AM: GROY

goldroyalty.com

David Garofalo and Amir Adnani

Gold Royalty Corporation (GRC) was founded in the summer of 2020 by GoldMining (New York Stock Exchange American [NYSE AM]: GLDG; Toronto Stock Exchange [TSX]: GOLD) and led through a successful initial public offering in early 2021 by founding CEO David Garofalo and founding Director Amir Adnani.

Garofalo has worked in various leadership capacities in the natural resources sector over the last 30 years. Prior to joining GRC, he served as president, CEO, and director of Goldcorp, a gold production company headquartered in Vancouver, Canada, until its sale to Newmont Corporation. Prior to that, he served as president, CEO, and director of Hudbay Minerals, where he presided over that company's emergence as a leading base metals producer. Previously, he held various senior executive positions with mining companies, including senior vice president, finance and chief financial officer (CFO), and director of Agnico Eagle Mines Limited, and as treasurer and other various finance roles with Inmet Mining Corporation from 1990 to 1998. He was named Mining Person of the Year by *The Northern Miner* in 2012 and Canada's Chief Financial Officer of the Year by Financial Executives International Canada in 2009.

Adnani is an entrepreneur and the chairman of GoldMining, where he directed the growth of a gold resources acquisition and development company with a sizable portfolio of gold projects across the Americas. Adnani is also the president, CEO, and a founder of Uranium Energy Corporation. Here, he advanced the company from concept to U.S. production in its first five years. He is chairman of Uranium Royalty Corporation, a company listed on the Toronto Stock Exchange Venture Exchange (TSXV) and NASDAQ Capital Market as a uranium royalty and streaming company.

The creation of GRC came about through the fortuitous introduction of Garofalo to Adnani in the fall of 2019. Garofalo had recently completed the sale of Goldcorp to Newmont, and Adnani was going through the process of an IPO for Uranium Royalty Corporation.

Through their conversation, Adnani noted that he was also the chairman of GoldMining, a company with 14 exploration- and development-stage gold and gold-copper projects throughout the Americas with more than 32 million ounces of global gold equivalent resources. The GoldMining team had assembled these assets through the bottom of the gold cycle between 2012 and 2019. As the gold market was starting to turn in 2019, Adnani was looking for ways to unlock the value of the significant resource base that GoldMining had accumulated in its portfolio. Through our discussions, we came to the idea of creating a new royalty company with a portfolio of royalties written on the GoldMining portfolio.

To this end, GRC was created with a portfolio of 14 net smelter return (NSR) royalties ranging from 0.5% to 2.0%. A management team and board were assembled with more than 300 years of collective experience in senior mining roles. On the management side, Garofalo took the role of president, chairman, and CEO. John Griffith joined as chief development officer, Josephine Man as CFO, and Alastair Still as director of technical services. On the board were Adnani plus Warren Gilman, Alan Hair, Garnet Dawson, and Ken Robertson. Ian Telfer, former chairman of Goldcorp and the World Gold Council, also joined as chairman of the gold royalty advisory board. With this world-class team and foundational development-stage portfolio, GRC completed a successful $90 million IPO on the NYSE AM in March 2021.

In connection with the creation of GRC and the initial IPO, GoldMining would also shift its strategy from being an acquirer to focusing on advancing and otherwise unlocking the value of the projects within the portfolio. Still was appointed the CEO of GoldMining in April 2021 while simultaneously holding the role of director of technical services at GRC. Still is a professional geologist with decades of operating and capital markets experience, most recently as the director of corporate development at Newmont and GoldCorp. By advancing the GoldMining projects under Still's leadership, the value of the associated royalties would increase. Given that GoldMining was, and still is, GRC's largest shareholder, this provided a virtuous cycle for both companies. Since 2021, the technical team has grown, and several projects of the GoldMining portfolio have been advanced, including the Whistler (United States), Sao Jorge (Brazil), and La Mina (Colombia) projects.

When looking at the universe of royalty companies in 2019, we had two key takeaways:

1. The royalty companies trade at premium price-to-net-asset-value (P/NAV) multiples and the largest royalty and streaming companies enjoy the highest multiples.
2. The royalty space was crowded, with too many small players, and ripe for consolidation. To effectively compete with the largest royalty companies, GRC needed to quickly achieve scale.

GRC's strategy was to consolidate the small cap royalty companies to create a company with critical mass to have capital market relevance and achieve a P/NAV multiple afforded to the seniors: Franco-Nevada Corporation, Wheaton Precious Metals Corp., and Royal Gold (Figure 1).

Below these three blue-chip royalty companies, all others are effectively small cap when viewed from the lens of the broader capital markets, that is, less than $2 billion in market cap, which gives limited trading liquidity and a lack of scale to compete with the seniors. As a result, all other companies within the royalty and streaming space trade around 1.0× net asset value (NAV).

Following the GRC IPO, the GRC team immediately started consolidating the sector with the acquisition of Ely Gold Royalties in August 2021. Ely brought 90 royalties, primarily focused in Nevada, which diversified the initial

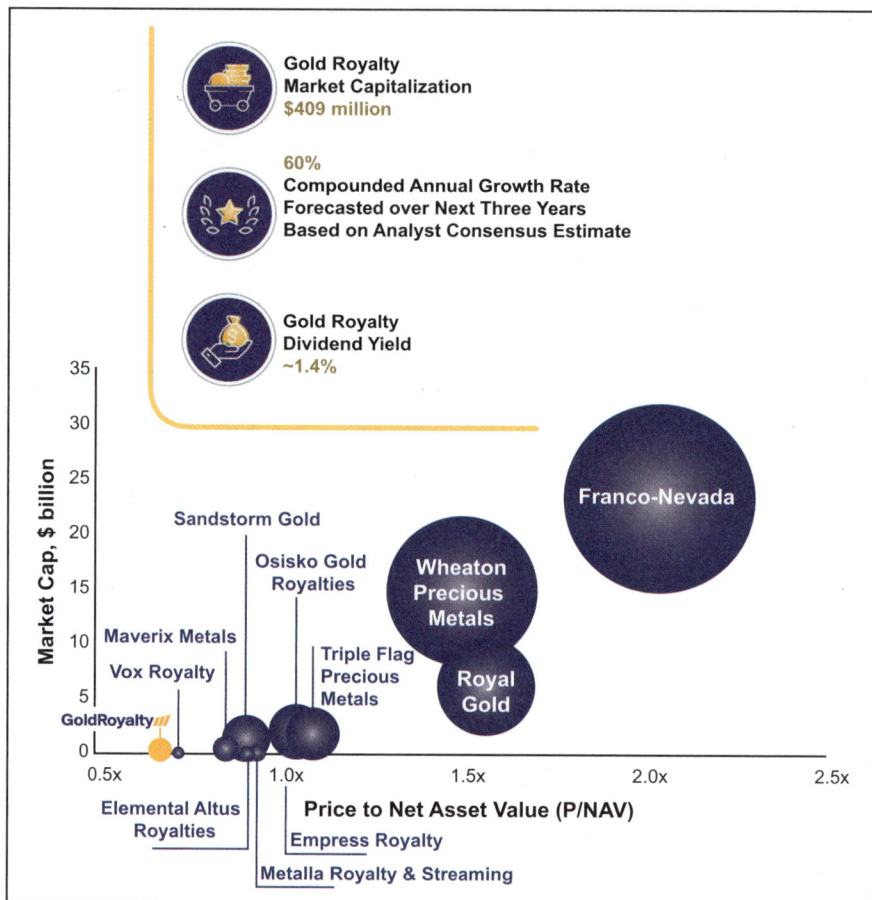

Based on peer group data from CapIQ, October 2022

Figure 1 Gold Royalty has significant room for growth in the sector

GRC portfolio and brought cash flow to the initial development-stage assets. Ely Gold Royalties had been created by Trey Wasser and Jerry Baughman who then joined GRC, with Wasser joining the board and Baughman joining the management team.

Following the Ely Gold Royalties acquisition, GRC completed the acquisition of Golden Valley Mines and Royalties and Abitibi Royalties. This brought another 90 royalties to the portfolio, including the company's cornerstone asset, a 3% NSR over the Odyssey underground project (Québec, Canada), the underground extension of the Canadian Malartic mine (Québec). Similar to the Ely transaction, Glenn Mullan, the founder of Abitibi Royalties and Golden Valley Mines and Royalties joined the GRC board.

Through the three companies acquired, GRC had grown the initial portfolio of 14 royalties to more than 190 royalties in just 12 months. This growth rate was unprecedented in the royalty space, and its consolidation strategy truly disrupted the space. From 2015 to 2020, royalty companies were focused on growth through incremental royalty acquisitions, and there was an absence of a true consolidator. With GRC entering the space in 2021, the music of consolidation had started. One of GRC's peers said, "You guys [Gold Royalty], really kicked over the ant's nest." Currently, Sandstorm Gold's acquisition of Nomad Royalty Company has just been announced, and the theme of consolidation is being adopted by other royalty companies in the sector.

The original portfolio was written on the 14 GoldMining assets located throughout the Americas. Through the various acquisitions that GRC completed in its first year of existence, we were able to build a long-life portfolio with royalties over three of the largest gold mines in North America:

1. Canadian Malartic (Québec, Canada). Through the acquisition of Abitibi Royalties, GRC acquired a 3% NSR over the Odyssey underground project at Canadian Malartic. The project is operated by the Canadian Malartic partnership owned by Agnico Eagle and Yamana Gold. Canadian Malartic is Canada's largest gold mine, and Odyssey is scheduled to begin production in 2023 with a multidecade mine life, providing meaningful, long-term cash flows to the company. The project ramps up production toward 2027 through 2029 and is expected to provide roughly 550,000 ounces of annual gold production.

2. Côté (Ontario, Canada). GRC had acquired a 0.75% NSR on IAMGOLD Corporation's Côté gold project, a development-stage asset scheduled to begin production in early 2024, which will become Canada's second largest gold mine with roughly 500,000 ounces of annual production over the first five years of the mine plan.

3. REN (Nevada, United States). Through the acquisition of Ely Gold Royalties, GRC acquired a 1.5% NSR and 3.5% net proceeds interest over the REN project, which is the underground extension of the Goldstrike mine. The project is owned by the Barrick Gold/Newmont joint venture, Nevada Gold Mines, and is the largest gold mine in the United States with more than 60 million ounces of historical production. The REN project had a maiden resource published in early 2022 outlining a 1.25 million ounces mineral resource at a grade of 7.3 grams gold per metric ton. Barrick CEO Mark Bristow has highlighted the project as being key to the near-term advancement of the growth of the Carlin complex. The project remains open in all directions with similar grades found over 2,000 meters from the edge of the current mineral resource.

By April 2022, across the balance of the portfolio, over 75% of the company's NAV was in Nevada, Québec, and Ontario.

LESSONS LEARNED: SEGUELA (CÔTE D'IVOIRE)

Prior to GRC's IPO, the company entered an agreement to acquire a royalty over Roxgold's (since acquired by Fortuna Silver Mines) Seguela gold project in Côte d'Ivoire, one of the highest-grade open pit projects in West Africa. The company entered into an agreement to acquire the royalty for about $16 million from the third-party royalty holder, Apollo Consolidated.

The acquisition was subject to a right of first refusal (ROFR) retained by the operator, Roxgold. As Roxgold was actively drilling the project and preparing for construction, we felt they were not in a place to pay up to exercise the ROFR and extinguish the royalty. We knew the risk but took a calculated bet.

Franco-Nevada, the largest royalty company in the world, proposed a restructuring to Roxgold, who decreased the royalty by 50%, with Franco-Nevada effectively buying the restructured Sequela royalty for $8 million.

What this highlighted to us is that the space is incredibly competitive with even the largest companies acquiring royalties that are insignificant for their purposes. This emphasized the importance of achieving scale quickly.

MERGERS AND ACQUISITIONS TRACK RECORD

In looking at the mergers and acquisition of Ely Gold Royalties, Abitibi Royalties, and Golden Valley Mines and Royalties, we learned that every deal that doesn't come to fruition always fails because of social reasons. With the successful acquisitions of these three royalty companies in 2021, GRC not only acquired a portfolio of top-tier assets but also retained the intellectual capital associated with those companies. These deals happened because GRC provided a compelling valuation to management, and all the parties believed in the objective of consolidation and the value that would be created for all shareholders through the strategy. As previously mentioned, Mullan, Baughman, and Wasser all joined the board or management groups and brought a new dimension to the GRC business model: royalty generation.

Baughman and Mullan are both prolific prospectors in Nevada and the Abitibi region, respectively. Their business model is to stake claims on exploration properties at a minimal cost and then vend these properties to explorers and developers. In return for the property, GRC receives option payments as a purchase price and also retains a work commitment and royalty by the operator. More than 40 of the assets within the GRC portfolio were created using this model. GRC gets paid to create royalties using this model rather than paying up for royalties in competitive processes.

Given the company is still less than two years old, the strategy remains unchanged, and the royalty market is still too crowded. However, there is plenty of room for consolidation in the sector, and all the small cap companies should be consolidated down to two companies. These companies would have meaningful scale and could strive to compete with the seniors for meaningful streams.

11

MAVERIX METALS

TSX | NYSE AM: MMX

maverixmetals.com

Dan O'Flaherty

The Maverix Metals story does not begin with "once upon a time" or "in a galaxy far, far, away," but rather, the original concept of creating a royalty company came directly from our Chairman Geoff Burns and ultimately culminated in the listing of Maverix on the Toronto Stock Exchange Venture Exchange (TSXV) in July 2016. The founding of Maverix was based on the simple premise of surfacing value for Pan American Silver Corporation in respect of a small collection of royalties that received no market value buried behind a meaningful portfolio of operating mines. Maverix's creation was not just an opportunity to surface immediate value for Pan American, but also to benefit from significant ongoing exposure to a new and growing royalty company.

In February 2015, Pan American announced that Michael Steinmann would be promoted to the role of president (which had previously been held by Burns in addition to his role as CEO). This announcement essentially signaled to the market the transition plan for Steinmann to take over as CEO and president, which he did effective at the end of 2015.

Throughout 2015, while Burns was capably managing a smooth transition with Steinmann, he and I had several conversations regarding where he could potentially continue to add value in the mining industry, given his extensive operating experience. Previously, Burns and I had worked on a potential transaction in 2013 that had several appealing elements, including the concept of surfacing value from assets buried in Pan American's operating portfolio. The transaction was one that Pan American had announced with Esperanza Resources Limited, whereby Pan American would vend-in three small non-core gold assets to Esperanza and take back a meaningful equity ownership position in a diversified gold development company. At the time, I was in corporate development at Esperanza, working closely with the Pan American team. Ultimately, the transaction did not complete, but the concept of surfacing value from non-core assets resonated with both Burns and Pan American.

From 2013 to 2015, precious metals equity markets were languishing. Revisiting the previous idea of spinning out small non-core gold development assets into a stand-alone company with no cash flow and with meaningful capital needs was simply a nonstarter. The equity markets were just not there to support a vehicle like that.

We quickly realized that we had simply arrived at the starting line.

As Burns continued to contemplate post-retirement, it became evident to him that Pan American had collected a modest, but valuable, portfolio of royalties through various land swaps, joint ventures, asset sales, and mergers and acquisitions (M&A). Knowing that I had extensive experience working on royalty and streaming transactions, Burns bounced the idea off me. That simple conceptual conversation led us down a lengthy path toward creating Maverix. I admit that when Burns first specifically mentioned royalties, I did not even realize Pan American owned any. Burns's response was that he did not think everyone internally realized that either, so most certainly the market would not be giving them any credit or value whatsoever. Given the transition of the CEO role to Steinmann, Burns had opportunity and interest in stewarding Maverix as chairman, which was critical in providing comfort to Pan American that the new vehicle would come out of the gate with proven leadership.

While the concept of launching Maverix was relatively simple, the execution was far from it. The initial step was incorporating a private company. We began with just Burns and me as shareholders, and then proceeded with identifying potential independent directors, public listing options, and planning out what the surrounding team and infrastructure would be comprised of. We then completed our valuation work, and initial thoughts on a proposed structure to present a fully baked plan to Pan American.

The first and most obvious hurdle was the related-party sensitivity of Pan American Silver potentially considering a transaction with its former CEO as the chairman of the counterparty. With the very (appropriately) high corporate governance standards of Pan American, this required legal bills, a special committee (including some interestingly tense negotiations between Burns and his former fellow directors), an independent valuation from a large investment bank, legal bills, and a few more legal bills.

Another notable obstacle was the underlying royalty portfolio itself. The portfolio consisted of a few royalties on development-stage projects in the Americas with minimal cash flow. Creating a separate vehicle with minimal cash flow in a very challenging financing market didn't feel like the most obvious path to success. Burns then had the idea of adding a minor gold stream on Pan American's La Colorada silver-lead-zinc-gold mine in Mexico, and one of Pan American's flagship assets. It was not necessarily obvious at the time as to why Pan American would burden one of its own assets even though the minor by-product stream had negligible impact on the underlying asset (please note the negotiations I referred to in the previous paragraph). But it was this addition that proved absolutely essential to the success of Maverix's launch. The cash flow from the La Colorada stream provided Maverix with a cornerstone asset out of the gate and gave much needed critical mass to maximize the value of the entire portfolio for all shareholders, including Pan American.

The discussions and negotiations with Pan American resulted in a formal agreement and announcement in April 2016. The deal closed a few months later, and Maverix began trading on July 11, 2016. On close, Pan American was Maverix's major shareholder at 63% and effectively was also our land-lord, given that we were renting a tiny sliver of extra space in Pan American's offices. Getting to the initial listing in July 2016 certainly felt like we had gotten to the finish line, although we quickly realized that we had simply arrived at the starting line.

While we had been busy beavering away at getting the company listed, the sentiment in the gold price and gold equities was starting to change. Specifically, in early 2016, as concept was turning into draft proposals, the gold price was hovering just below $1,100 per ounce. To say that precious metals sentiment was negative at the time would have been an understatement. Gold had been consistently trending downward from peaks in 2011 and 2012 in the $1,750 per ounce range. So, it had been several down years after the previous extended bull market from 2003 through 2012. It seemed to be an almost ideal time to start a royalty company and start accumulating assets while there was generally no financing available and plentiful opportunities to deploy capital. The royalty model, combined with a very low general and administrative cost approach and modest cash flow put Maverix in a position where it could hunt for opportunities and not be in any financial jeopardy. While we had restrained cash flow and limited cash on hand, we were confident that if we identified and evaluated high-quality projects with high-quality teams, we would be able to source capital.

However, by the time the company was listed in July, the gold price had shot up from less than $1,100 per ounce to $1,350 per ounce. We had several internal conversations wondering and somewhat bemoaning whether or not we had already missed the window. As readers of this book are aware, the number of royalty and streaming opportunities can sometimes be counter-cyclical to the availability of alternative forms of financing, such as equity markets. Fortunately, and surprising to some, there tend to be royalty and streaming opportunities in all metal price environments. Specifically, in lower price environments, there tend to be more debt restructuring financings, a focus on minimizing dilution in depressed pricing environments, and royalty financings being part of M&A solutions. In rising metal price environments, projects look better economically, executives and investors are more excited about moving those projects forward, and, simply, more projects need financing. That is just on the precious metals side, as there is an additional domain of opportunities searching for precious metal by-product streams from base metal producers that can have different price cycles.

We had a clear vision as to what types of royalty and streaming assets we would like to build the portfolio around, without actually knowing the specifics as to how we would get there. The main criteria were actually relatively simplistic: predominantly precious metals, relatively low geopolitical risk, strong operating partners, and of course a technically robust underlying asset. Initially, we also had a focus on producing assets or assets with a clear path to production to build up a "cash flow engine" that we could use to redeploy into future growth opportunities. We also wanted to create a simple-to-understand story for investors. We were committed to being a pure play royalty company without any direct mining interests or other bells and whistles that might be difficult for investors to value or understand.

Interestingly, and unexpected to us at the time, the most fertile ground for Maverix ended up being pre-existing royalties that were becoming more valuable as gold prices increased and projects that had previously been shelved were starting to move forward again. We would like to pretend that perhaps that was the grand strategy the entire time, but entering that target-rich environment was more luck, timing, and having the platform in place to act quickly, than an intentional strategy.

Before concluding any M&A though, we did complete a round of initial marketing to introduce Maverix. It is probably fair to say that we were proud of the work that it took to get to launch and were excited to tell the story, but the initial reception was dismal to say the least. While the gold price had done well that year, there was not a lot of excitement for a tightly held, small, junior royalty company with no track record of external deal flow. In fact, the reception was so muted that one fund manager in Toronto quipped "What do we need another royalty company for?"

In retrospect, this question seems much funnier now. At the time, the answer to that question seemed obvious. The public precious metals royalty sector landscape was quite different then, with the senior most companies being Franco-Nevada Corporation, Wheaton Precious Metals Corp. (formerly Silver Wheaton Corp.), and Royal Gold, intermediates being Sandstorm Gold, and several smaller juniors that had more of a prospect generator model. At that time, Osisko Gold Royalties had a market cap of approximately $1.3 billion, Sandstorm approximately $600 million, and Triple Flag Precious Metals was in the process of being created as a private vehicle, so there was still a sizable gap to Maverix's initial market cap of around $60 million. While we did not know what the exact specifics would be, it seemed to us that there would be a very significant opportunity set of smaller royalty and streaming transactions that would be material to Maverix but wouldn't necessarily move the needle for the much larger players. It appeared to be an under-focused transaction size area, and our strategy was to compete below the radar of the larger players. Ultimately, Maverix ended up being one of the first movers into the smaller end of the royalty sector as a plethora of new entrants have emerged since Maverix was launched in 2016.

After an eye-opening round of initial marketing, we knew that we had a lot of work to do to build the underlying business before we could attract any investor attention. Fortunately, this came together through an introduction to Gold Fields Limited, specifically to Johan Pool in the Denver office. Gold Fields had accumulated a small portfolio of royalties itself, similar in value to the Maverix portfolio but more weighted toward near-term cash flow. At the time, Maverix owned 13 royalties and streams, two of which were paying, but predominantly development assets in the Americas, while Gold Fields owned 11 royalties, four of which were paying, predominantly in Australia, and headlined by a 2.5% net smelter return (NSR) royalty on the Mt. Carlton mine (Queensland, Australia; then operated by Evolution Mining). The portfolios were highly complementary in nature. Gold Fields had indicated to us that they had evaluated scenarios of putting their royalties into a public vehicle and retaining equity to participate in a re-rate of a growing royalty company. The concept was simple in that critical mass in the royalty sector is particularly important to valuation, and there was an opportunity for two plus two to equal five.

Over the course of the next several months, it was the primary focus of Maverix to work with Gold Fields, leading to the successful announcement in December 2016 of the acquisition of a portfolio of royalties and the welcoming of a new core shareholder to the registry. This transaction effectively doubled the size of the company, significantly increased immediate cash flow, and brought a new supportive major shareholder. The transaction was entirely for equity in Maverix (common shares and common share purchase warrants), and upon transaction close, it resulted in Pan American owning approximately 40% and Gold Fields owning 32%. One of the challenges going into the Gold Fields transaction was that Maverix was an unknown quantity to Gold Fields, and, by taking equity, they would have commercial exposure to the platform going forward. As the transaction progressed, we were able to build a strong and positive relationship and having the previous endorsement of Pan American was certainly an added benefit.

> *Wouldn't it make more sense to focus on building the best underlying business and let the liquidity grow naturally?*

The market reacted incredibly favorably to the Gold Fields transaction. Maverix enjoyed a positive re-rate of over 80% in the weeks following announcement of the transaction. This immediately proved the benefits of increased scale, diversification, and capital markets profile.

Our next round of marketing following the Gold Fields transaction was notably different. The transaction was positively received as was the proof of concept that Maverix would be able to grow going forward. We did, however, quickly run into another issue that would stick with Maverix for several years—liquidity. Given the incredibly tight float, Maverix was trading by appointment on thin volumes at the time. There was some modest interest in the story, but there was not much stock available for many investors to purchase. A bigger issue was on the institutional front where investors were more worried about how they would eventually exit if they ever needed to sell.

As we were contemplating ways to address the liquidity issue, a chance to simply ask was available, and I learned a great lesson from a conversation that Burns and I had with Ross Beaty, who was then chairman of Pan American and a very successful serial entrepreneur. We had been pitched by a couple of boutique investment dealers to pursue a small equity raise for the sole purpose of placing shares into the market. When we presented this to Beaty, the response was something along the lines of, "You want to dilute all your existing shareholders with absolutely no use of proceeds. Wouldn't it make more sense to focus on building the best underlying business and let the liquidity grow naturally?"

We have most definitely adopted that mantra at Maverix. We did not do an equity deal then, and in fact, six years later have still not completed a public equity offering from treasury. When we evaluate transactions, we always do so on per share metrics. As evidenced by the re-rating after the Gold Fields acquisition, critical mass is important, but not at the expense of just growth for the sake of growth regardless of per share metrics.

In 2017, we set off to continue to grow the business on the newly expanded platform. We did this through acquiring four cash flowing royalties in three separate acquisitions. In total, we deployed more than $40 million to acquire paying royalties on the Florida Canyon (United States), Beta Hunt (Australia), Silvertip (Canada), and Karma (Burkina Faso) mines. Some attributes that attracted us to these opportunities were that they were all-cash flowing, early on in their respective mine lives with a full mine life ahead of them, and generally in mining-friendly jurisdictions such as Nevada, Australia, and Canada. These acquisitions meaningfully increased our cash flow, but also our cash flow per share. The deals also established a track record of Maverix being able to acquire royalties independent of larger portfolios.

Going into 2018, the entire background to date set the stage for Maverix's largest and most transformational transaction. Newmont Corporation had been privately marketing a portfolio of 51 royalties, headlined by a 1% NSR on Hope Bay in northern Canada. It was an attractive portfolio with significant scale and assets that covered the entire production and development curve from operating assets to near-term development assets, exploration stage assets, as well as large out-of-the-money resource projects. We knew (correctly) that there would be significant competition as portfolios such as this rarely came available.

The scale of the transaction of approximately $100 million was considerable relative to our market cap at the time of approximately $175 million. Given that and the expected competition, we considered ourselves an underdog during the process, but we were keen to put our best possible foot forward. We believed the biggest differentiator in our proposal, for better or for worse, was the sizable equity portion of consideration versus an all-cash offer. Fortunately, we were able to build a great relationship with Newmont, specifically with Blake Rhodes, at the time Newmont's vice president of corporate development, who was tasked with managing the process and is now a member of the Maverix board. During the entire process, we were able to articulate the potential upside benefits of retaining equity and that the increased scale and diversification would be a key factor to reach higher valuations and continue to grow the capital markets profile. It was also helpful that the positive re-rate Maverix received from the Gold Fields transaction had already provided a proof of concept.

As the original endorsement from Pan American benefited Maverix in the Gold Fields transaction, having two large mining companies already on the share registry proved to be another competitive advantage.

In the second quarter of 2018, we were able to consummate this transaction for $17 million in cash, 60 million common shares of Maverix, and 10 million common share purchase warrants, welcoming Newmont as our largest shareholder. This transaction was a defining moment for Maverix as it transformed the portfolio from 27 royalties and streams to 78 and provided a greatly enlarged platform to continue to build on.

Much like the Gold Fields transaction, this transaction was very positively received by the market and within two months following the announcement, Maverix's share price increased by over 50%. With Newmont taking a considerable amount of equity consideration, the transaction seemed to be a true win-win deal.

Given the positive reaction from the market, it once again shone a spotlight on the lack of liquidity. Again, we had never (and still have not) done a public equity offering from treasury. Our shareholder registry appeared tight with the shareholder registry at the time: Newmont (28%), Pan American (26%), and Gold Fields (20%). It did not take a lot of math to realize the float was still very constrained.

At this point, we had achieved a level of critical mass in a time frame beyond even our own most optimistic assumptions, but we knew that we still had to build a liquid trade to increase the investor appeal. We began preparing all the documentation required to move from the TSXV to the TSX as well a formal application to list on the New York Stock Exchange American (NYSE AM).

The critical factor is having the right team in place.

Over the course of preparing the up-tier of our public listings, in early 2019, Gold Fields approached us about the potential to liquidate their position. The gold price at the time was in the $1,300 per ounce range, but there were not many notable equity financings in the precious metals sector. Additionally, with Gold Fields owning just under 20%, it represented a meaningful portion of the company. Despite the challenging markets, with the benefit of working directly with Gold Fields and prospective investors, we were able to place the entire position through a series of off-market block trades. Having been a shareholder for over two years and enjoying an increase of 150% (approximately a $60 million gain) on their position, Gold Fields was able to successfully crystallize that value and their final realized value on their royalty portfolio appeared to be much higher than it could have received had Gold Fields simply sold the portfolio for cash.

A significant positive for Maverix was being able to demonstrate that institutional demand actually existed versus retail shareholders or the major mining companies that had been our M&A counterparties, which gave us confidence heading into our NYSE AM listing.

Outside all the entire legal process and rigmarole, a pesky wrinkle of qualifying for the NYSE AM is a minimum $2 share price, which Maverix was well below. This required a share consolidation simply to meet the listing requirements. It was on this announcement that we received our first set of angry emails from investors demanding explanation on the share consolidation. While the process of a share consolidation is just math, they often have had a connotation of desperation or a harbinger of bad news (which was not the case with Maverix).

We ended up listing on the NYSE AM and moved from the TSXV to the TSX in mid-2019. Having the opportunity to ring the opening bell of the NYSE was absolutely a highlight for me, not just a Maverix highlight but an entire career highlight, particularly being able to share it with our team. The listing proved to be invaluable as it represented an inflection point on our liquidity, which has steadily grown with the bulk of which on the NYSE AM. Actually, not long after this, we opened a satellite office in New York, as Ryan McIntyre joined us from Tocqueville Gold Fund as president of Maverix.

Outside the portfolio acquisitions that had built most of the Maverix portfolio, we did have opportunities to deploy capital directly into operating mines when we made direct stream investments into the operators of the Moss mine in Arizona (United States) and the El Mochito mine in Honduras. This represented a slight departure from our modus operandi to date, and the process of creating a new stream versus buying pre-existing royalties was a very different, specifically, direct boots-on-ground site access, direct discussions with the underlying management team, and the ability to completely customize the streaming contract for both parties.

In 2019, the gold price was increasing rapidly from the $1,200 per ounce range to the $1,500 per ounce range, which presented challenges in pricing transactions for both us and any potential counterparties. During this period, we were invited to participate in a process that Kinross Gold Corporation was running on a portfolio of 25 precious metals royalties. The portfolio appeared to be a natural fit for Maverix, and we determinedly pursued the opportunity. As in past transactions, the support of our large shareholders in Newmont of Pan American was an advantage. Additionally, the ability of Gold Fields to successfully exit their position at a gain demonstrated a capacity to crystallize a sizable equity block. In the fourth quarter of 2019, we announced the acquisition of the portfolio for approximately $74 million, comprised of $25 million in cash and the balance in Maverix equity, and we happily welcomed Kinross onto the registry as yet another major mining company shareholder.

Entering 2020, the royalty sector was starting to see several new entrants, particularly on the junior side. As well, several of the larger royalty companies were starting to expand their size criteria downward (e.g., evaluating smaller opportunities in addition to their traditional hunting grounds). This squeeze started to put pricing tension on numerous royalty transactions, and we began to see an erosion in transaction returns. Instead of chasing unsustainable financial returns, we looked to our existing deep relationships for win-win transactions. In 2020 and 2021, we acquired a second portfolio from Newmont, a second portfolio from Pan American, and a second transaction with Auramet Trading.

As of this writing, in mid-2022, Maverix recently hit its sixth anniversary as a public company and has grown from an initial base of 13 royalties and streams, two of which were paying, to 125 royalties and streams, 14 of which are paying, through a series of 14 separate transactions. This growth profile highlights one of the benefits of the royalty model in general, in that it can be scalable with a relatively small team. Maverix itself currently has nine employees with an ability to scale up the asset portfolio without necessarily corresponding increases to general and administrative costs. The critical factor is having the right team in place and this is likely true of many of the successful companies referred to in this book. We certainly could not have gotten very far as a business venture without the people that we have in place as Maverix employees, board members, and shareholders, as well as our operating partners, and advisers in all capacities. There are not enough thanks to go around for everyone involved, but their contributions are very much recognized and noteworthy. Onward and upward.

12

TRIPLE FLAG PRECIOUS METALS

TSX | NYSE: TFPM

tripleflagpm.com

Shaun Usmar

PRIVATE EQUITY RAISES THE TRIPLE FLAG

Toward the end of 2015, one of the precious metals sector's biggest names, Barrick Gold Corporation, was emerging from a difficult but transformative year. The previously debt-laden group had been slimmed down, and its share price had staged a dramatic recovery. Gold and silver prices were still hovering around five-year lows, but Barrick was back on track.

As Barrick's chief financial officer (CFO) at the time, I was one of the architects of that turnaround. With the recovery at Barrick well underway, most investors expected me to spend the coming 12 months bedding in at Barrick, but in April 2016, I left Barrick to launch Triple Flag Precious Metals Corp., stepping away from a career within mining companies and into the world of streaming and royalty financing to partner with them.

The launch of Triple Flag Precious Metals was not just a new direction for me; it also marked a new chapter in the streaming and royalty business—the arrival of global private equity.

A PRIVATE PROPOSAL

I was preparing for a Barrick investor day at the New York Stock Exchange and the publication of end-of-year financials when I got the call from a hedge fund in New York about an opportunity. At first, I politely declined the approach, but they were persistent, and I could not quite get the idea out of my mind.

I never want to look back on things and think, "Damn, I wish I'd tried that." So, I followed my curiosity, took a day's leave, and flew to New York. I'll be honest, I came at it with a biased perspective. I thought I would be meeting with some overly confident New York bankers, but, instead, the people I met with were sincere, engaging, and smart. Four hours of conversation passed by in a moment.

The financiers were from Elliott Investment Management L.P., and the proposal was to launch a streaming and royalty business from scratch, with a focus on precious metals. The proposed backing from Elliott would be $1 billion, with a view to investing roughly $200 million a year in streaming and royalty deals alongside the opportunity to build a team and business from scratch, which I found compelling. The opportunity reminded me of the early days of Xstrata, where I was a founding senior executive and loved the energy and challenge of building a business with talented people.

I already had experience with streaming and royalty deals at Barrick. In fact, a feature of the Barrick turnaround in 2015 had been a $610 million streaming deal with Royal Gold. This deal informed my views of the merits of the streaming and royalty business model relative to other traditional forms of financing that CFOs of producing mining companies had to choose from. It also provided insight into the more rational and contained competitive dynamics among the larger, more sophisticated streaming companies that I aimed to directly compete with, utilizing Elliott's firepower and the capabilities of the team I planned to build. That deal, and my experience in base metals mining companies like BHP and Xstrata, had demonstrated to me that there were win-wins to be had through streaming, particularly when partnering with base metals businesses with precious metals by-products that weren't particularly valuable within a base metals or diversified mining company. These precious metals by-products offered a clear value arbitrage for a precious metals investor that coveted these ounces and valued them more highly. Streaming also offered a more patient and customizable form of financing, with favorable risk characteristics and competitive funding costs. For a sector that often lacked access to traditional forms of debt and equity financing at key moments in the commodity cycle, well-capitalized streamers could provide alternative forms of funding when it was needed most. Different players had different cash flow needs, and streaming and royalty financing could unlock value to parties on both sides of the transaction.

Key to the Triple Flag plan was the time horizon on which Elliott was prepared to work; it was private capital, but, most importantly, it was patient evergreen capital without a set investment horizon—quite unique in the resource private equity space. This evergreen time horizon was one of two key factors that were crucial to my decision to start the business and to Triple Flag's success. The other factor was the team I was able to assemble, comprised of experienced mining and financial professionals. While at Xstrata and Barrick, I sat on the other side of the table in deal negotiations, and, along with the wider Triple Flag team, our experience in the executive and operational side of mining and exploration as well as deal-making spanned nearly two decades.

While the capital was available for investment, Triple Flag began like any other start-up: finding office space, deciding on a name, building a website on the cheap, and buying office furniture for cents on the dollar. But the real work of searching for suitable assets began immediately. Triple Flag examined many potential deals during its first months. There was just one two-week period when there was not a prospective deal under consideration. It was an anxious two weeks—the only calm period in our six-year journey. However, prospects returned, and, toward the end of 2016, the first deal was coming to fruition.

CLINCHING THE FIRST DEAL

Triple Flag took the proposal to Elliott's chief executive, Paul Singer. We hadn't really met Singer at that point, and he began the meeting with the old Mark Twain quote: "A mine is a hole in the ground with a liar standing at the top." Given the mining sector's reputation for being overly promotional and destroying shareholder value in prior cycles, it wasn't an unusual sentiment.

I replied, "If that's the benchmark, I think we can exceed your expectations."

Triple Flag's backers were deeply analytical and attuned to the risks, optionality, and potential for asymmetric returns. But, as well as providing patient capital, Elliott was confident it had backed the right business. After a series of rigorous investment committee meetings as a new and untested team, we had our first binding mandate—a key milestone.

The deal in question was indeed a lot better than just a hole in the ground. It was a $250 million silver streaming agreement with Compañía Minera Milpo covering its fully operational Cerro Lindo zinc mine in Peru. The site was nearly 150 kilometers southeast of Lima on a private road, at an altitude of around 2,000 meters above sea level.

> *There are win-wins to be had through streaming, particularly when partnering with base metals companies with non-core precious metals by-products.*

On our first visit to the mine, Triple Flag used what would become our now proven approach to conducting rigorous due diligence, which was to supplement our experienced team of technical; environmental, social, and governance (ESG); and commercial professionals with specialized external experts in key disciplines, as needed, from the company's extensive global networks. I found this model to be highly effective when working in mergers and acquisitions at Billiton and Xstrata, enabling them to keep overheads low while allowing for rigorous review and agile decision-making.

I had excellent talent pools to choose from when building my core team; for example, I received more than 200 resumes when appointing a geologist and more than 400 for the most recent analyst hire. Rather than using headhunters, I depended on my and the team's knowledge and networks in the mining industry to source and evaluate candidates for the management team, and later for the compilation of a high-quality, independent board. The value of this hybrid talent model was on display early for the site visit to Cerro Lindo, where Triple Flag had tracked down the recently departed head of exploration at Milpo to join the due diligence team, offering deeper insights into the exploration potential and future optionality of the land package.

This very first deal vindicated the Triple Flag approach of using the best possible technical expertise for each unique situation and integrating the findings into a compelling commercial offering while utilizing astute deal tactics learned through many years of senior executive experience in major mining companies.

The bigger streaming companies have larger technical teams, but, by mining company standards, they are still small. I've worked in some of the largest mining companies in the world that employ hundreds of good technical people. However, what really matters is the quality and depth of experience of the people you rely on, and how you integrate the analysis from the various disciplines into rigorous investment proposals that showcase good judgment and decision-making, and I think we've demonstrated that on several occasions.

Just a mere seven months after I founded Triple Flag, we announced we had sealed the Cerro Lindo deal. The Cerro Lindo deal was also a model example of the Triple Flag strategy that was to follow—investing into long-life, polymetallic base metals mines with a stream on precious metals by-products that were not core to the operation.

BUILDING SUSTAINABLE GROWTH

The silver by-product streamed from the producing Cerro Lindo mine was immediately cash generative for Triple Flag. That really got us into the game and allowed us to build Triple Flag as a private player. Incentivizing the talent from the outset was also key, and the private equity model had allowed all members of the small management team to take a stake in the business.

I called on my friend and former colleague Mick Davis, the former chief executive of Xstrata, for periodic advice during the early days of the business, before Davis later agreed to join the company board. Davis was one of several mentors I relied on for advice at key moments while building the business. The career history of Triple Flag's management team and its contacts in some of the world's largest mining groups provided additional references and access to potential streaming and royalty opportunities.

Aside from that two-week period in its very early days when deal flow fell silent, Triple Flag has never been without a deal under consideration. During the first six years of its existence, we examined more than 600 opportunities and completed 19 transactions. Being highly selective and disciplined in our investment decisions has been key to our successful track record.

The focus has been on larger deals, typically in the range of $100 to $500 million, involving assets that are either already cash flowing or construction-ready, fully permitted projects that are ideally within two years of production. These deals are usually linked to large, prospective land packages that offer the potential for future exploration success and life extension beyond the visible mine life. Today, we have 80 assets across the Americas, Europe, and Asia: 15 in production that account for around 90% of the portfolio value, and 65 in development or exploration stages. We have maintained an overwhelming focus on precious metals, which account for over 80% of our streams and royalties.

Our approach has been to try to get in the mind of each operator to offer enabling funding solutions. You need to understand their needs and what they are trying to achieve to be able to offer competitive proposals. This approach also goes hand-in-hand with excellent ESG leadership when screening for good mining investment partners.

Increasingly, the question is: "What benefits can we offer to our mining partners beyond just the cost of capital?" In addition, these relationships extend for decades, so we have to know our counterparties are people who we want to be in business with for a very long time, and that they share our values on key issues related to ESG practices and maintaining their privilege to operate. We like to invest alongside them into local community initiatives near the mine or project to enhance their efforts where we can. This is just smart business. If you are at a crossroads with your local communities or your host governments, you cannot pick up your mine and move it somewhere else. It is also good practice to be a good custodian of these non-renewable resources and a good partner to the local communities that support the mine or project.

The cornerstones of Triple Flag's portfolio are:

- A gold and silver stream on CMOC Group Limited's Northparkes mine in Australia
- A silver stream on Nexa Resources' Cerro Lindo mine in Peru
- A gold stream on Royal Bafokeng Platinum (RBPlat)'s platinum group metals operations in South Africa
- A royalty on Agnico Eagle Mines Limited's Fosterville mine in Australia

Throughout its rapid growth, Triple Flag has maintained its start-up entrepreneurial culture. Our key strengths are our engaged culture of owner-managers in the business, our flat structure, our small and highly expert team, and our agility.

Simple things, which would take multiple levels of approval in other bureaucratic organizations, our people can just get on with and make empowered decisions with clear and delegated authority. We want to make a difference, so it's all about getting things done for the benefit of our partners and stakeholders alongside a team of people you respect and admire, who challenge each other to be better every day.

FROM PRIVATE TO PUBLIC—RUNNING THE IPO GAUNTLET

Triple Flag's agile approach was enabled by our private equity backing, allowing us to focus on sourcing and executing high-quality deals to build the portfolio. Triple Flag, though naturally well known to the industry, remained almost invisible to the wider market while private.

However, it was never the intention to remain a private equity business forever. At the outset, the plan was to likely pursue an IPO with a rough time frame of five years and, by 2019, the company was finding it hard to continue building the portfolio unnoticed. At the same time, precious metal prices, which had been volatile since 2016, were now rising. We had investment bankers increasingly coming to us saying, "You've had a good run. The demand for gold is going up, and we think you have an impressive track record and differentiated growth story. We think you're ready to go public." So, we were finally persuaded to give it a try.

The planned 2019 IPO, however, was not to be. Plans and a prospectus were drawn up, but late in that year, Triple Flag withdrew the public offering because of the volatile market conditions. We didn't need the proceeds, and Elliott remained fully supportive of the team and business, so we decided to maintain our discipline and reject the sub-optimal deal on offer, choosing instead to continue to grow and mature the portfolio. Less than a year later, Triple Flag secured its largest deal yet with the support of Elliott, a $550 million gold and silver stream on CMOC's Northparkes copper-gold mine in Australia. Then, in the latter half of 2020, Triple Flag revived the plans to go public, and in May 2021, an IPO was completed, raising just over $263 million and valuing the group at about $2 billion. It was the largest Toronto Stock Exchange (TSX)-listed mining IPO in a decade and the largest global precious metals IPO by market capitalization since 2008.

> *You need patient, sophisticated capital in combination with the talent and experience in the team to succeed.*

Going public brought some immediate pain. The stock had been nearly three times over-subscribed at IPO, and I thought that the stock would quickly trade well above IPO levels. But completion of the IPO coincided with talk of interest rate tightening by the U.S. Federal Reserve, and the price of gold dipped. The stock tracked sideways and then began to drop. The pain went on for five months before the price bottomed out and began a steady rise back up to, and beyond, its IPO price.

Those were hard months. We were doing what we said we'd do, we were well-capitalized, and the business was performing well, but there was this absolute torture happening in the markets that was divorced from the fundamentals.

Stepping out of pure private equity ownership had brought public scrutiny, but the transfer from private to public also inevitably changed the demands on the company. The group had maximized the benefits of its private equity start: a tight-knit senior team, a flat hierarchy, and the capacity for rapid and agile decision-making. Public company status brought new challenges, but the company's systems and governance were set up with public life in mind from the very beginning, and the essential simplicity of Triple Flag has pleasingly survived and remains a boon.

How do you deal with the regulatory and other burdens but maintain that sort of entrepreneurial agility that enabled our success? There are all these additional burdens associated with being public that are distracting and take time, money, and resources that, as a private company, are just less onerous. But our team is extremely capable and is working hard to deal with the transition, because a lot of what we are doing is setting up established processes for the first time that will become routine in the future.

On the other hand, I have the experience of working as an executive in very large public organizations—many tens of billions in market cap—and Triple Flag is by far a much simpler and more scalable business. So, the intrinsic character of the company and the essential fit of the team members are all alive and well. The leadership team also retains a substantial interest in the equity of the company, which is distinctive and quite unique among our public peers. We think like owners, because we are owners, with each team member participating in and having most of their net worth tied to the equity of the company.

The Triple Flag route of private equity to public offering was a significant innovation in the streaming and royalty industry and allowed the company to rapidly build a team and a portfolio to its liking and away from the public gaze. I believe it is an approach that could work again for a new entrant, but it requires the right private investors, the right approach to the business model, and placing a vital emphasis on experience, expertise, and fit.

I think it is a bit harder for traditional private equity to invest into the resources sector because they often have a defined and limiting investment horizon that may not work with the realities of the commodity cycle. That patient evergreen capital is a necessity for exposure to multiple price cycles, but I think it is out there. You are seeing some pension funds starting to venture into the streaming and royalty space, and they themselves have very long-term investment horizons and like the top-line exposure and optionality that the streaming and royalty business model offers.

You need to have patient, sophisticated capital in combination with the talent and experience in the team. You need both to succeed. I'm truly lucky to have had Elliott as our founding capital provider together with the exceptional team we've built these past six years. Our track record together tells our story best.

13

JUNIOR AND MID-TIER COMPANIES

Compiled by Douglas Silver

This chapter presents the other active industry participants. Most of these companies have been formed in the past five years, and each one has a unique perspective on how to build shareholder value. Each company was asked for their outlook on the industry in the next 5 to 10 years. The answers for those who responded are contained in their profile. All market capitalizations are as of December 31, 2021.

ALTUS STRATEGIES

AIM: ALS	TSXV: ALTS	OTCQX: ALTUF

altus-strategies.com

Market capitalization: $94 million

Altus Strategies was a mining royalty company focused on establishing a diversified portfolio of income-generating royalties, royalty rights on pre-production projects, and direct project interests. With a risk-diversified business model, short- and long-term income was generated and also provided investors with exposure to the potential multiple returns that could be produced from the discovery process. Altus grew its portfolio of royalties through organic royalty generation and the potential acquisition of royalties from third parties. The company's portfolio approach reduced risk exposure through commodity and geographic diversification by entering joint ventures with third parties on its discoveries. Shareholder capital was preserved for investing in the next discovery opportunity. The generated royalties were designed to yield sustainable long-term income for Altus.[1]

Postscript

On June 14, 2022, Altus Strategies announced that it had reached an agreement with Elemental Royalties Corporation for an all-share merger of equals. Altus shareholders receive 0.594 Elemental shares for each Altus share held. The combined companies are called *Elemental Altus Royalties Corporation.*[2]

DETERRA ROYALTIES LIMITED

ASX: DRR

deterraroyalties.com

Market capitalization: $1,634 million

Established as an independent company in 2020, Deterra Royalties Limited's principal activity is the management and growth of a portfolio of royalty assets across a range of commodities, primarily focused on bulks, base, and battery metals. Deterra's existing portfolio includes royalties held over its cornerstone asset Mining Area C in the Pilbara region of Western Australia, as well as five smaller royalties in Western Australia including Yoongarillup, Yalyalup, Wonnerup, Eneabba, and St. Ives.[3]

---

1. "Our Strategy," Altus Strategies, accessed October 10, 2021, www.altus-strategies .com.

2. Altus Strategies, "Altus Strategies and Elemental to Merge and Create Significantly Enlarged Mining Royalty Vehicle," press release, June 14, 2022.

3. "About Us," Deterra Royalties Limited, 2021, www.deterraroyalties.com.

ECORA RESOURCES (FORMERLY ANGLO PACIFIC GROUP)[4]

TSX | LSE: ECOR

ecora-resources.com

Market capitalization: $383 million

Ecora Resources is a royalty company focused on acquiring royalties and streams on commodities that are required for the decarbonization of the world's energy supply and consumption. The company also pursues sustainably produced commodities that are needed for creating a sustainable future. The current emphasis is on copper, nickel, iron ore, gold, cobalt, vanadium, metallurgical coal, and uranium.

ELECTRIC ROYALTIES

TSXV: ELEC   OTCQB: ELECF

electricroyalties.com

Market capitalization: $22 million

Electric Royalties is a cash-generating royalty company with a growing portfolio of 19 royalties, offering exposure to metals crucial for the electrification of the world's auto fleet and energy generation: lithium, copper, cobalt, manganese, tin, zinc, graphite, vanadium, and nickel. Electric Royalties has been public since June 2020 and has demonstrated success in utilizing first-mover advantage in the clean energy metals royalty space.

The impetus for creating Electric Royalties was to build a company with the highest odds of success over the next three to five years, via a corporate strategy that is focused on royalties as a business model and clean energy metals as the target market.

---

4. Anglo Pacific Group, "Change of Name to Ecora Resources," news release, October 5, 2022, www.ecora-resources.com.

Diversification is important to us, to get away from the traditional single-asset investment model in junior mining, which is plagued with a 15-year average development timeline, development and permitting risks, and dilution to shareholders along the development curve. The royalty model avoids all these pitfalls, including the need to account for capital costs, cost overruns, and holding costs.

Our goal is to stay diversified, not only across royalty assets but also on a relatively even basis across each clean energy commodity. That way, we offer investors lower risk and direct exposure to the clean energy transition through the very building blocks and raw materials that enable this reality. We are not hedging on any one commodity, evolving battery chemistries or supply-side shocks because of new displacing technologies.

With regard to the target market, no other royalty company is solely focused on the clean energy metals space. There are many established major players competing in the gold and silver space and largely fighting over a pipeline of assets that existed 10 years ago. We believe this has caused the cost of acquiring royalties in the gold and silver space to soar, making economics much less attractive than those we can attain with our suite of clean energy metal projects.

Electric Royalties' focus is where the future growth in the mining royalty space will come from, which we believe are the metals required for the global transition to clean energy. Metal demand growth rates across almost all of the metals required to upgrade the world's infrastructure to support a decarbonized global economy via electric vehicles, batteries, energy storage, and renewable energy, have projected double-digit demand growth annually for the next 10 to 20 years. Electric Royalties is currently the only royalty group focusing on the entire suite of metals required for the clean energy transition.

We are also unique in how we have structured transactions via co-investment, paying most of the acquisition costs in shares, and creating royalty options payable in shares. We are one of the few junior royalty groups with experience in creating new royalty structures. We rigorously evaluate each potential transaction in safe jurisdictions with local metal demand (thus shorter supply chains) and favor projects earlier in the development phase, where we see and capture all of the incredible growth in value while third parties progress them along at their own expense. However, we also prioritize cash-flowing royalties in parallel with a view of paying a dividend, which is a pivotal milestone for royalty companies.

With more than Can$150 million raised for development by operating partners of the assets that Electric Royalties holds royalties on, we're expecting major catalysts across most of our portfolio in 2022. With one producing royalty, we expect additional royalties coming into production over the next 12 to 24 months and project that our royalty portfolio could be paying out $20 million annually within five years—all at no additional costs to the company and no significant dilution to shareholders. We believe we're one of the most undervalued royalty companies.[5]

ELEMENTAL ALTUS ROYALTIES CORPORATION
(FORMERLY ELEMENTAL ROYALTIES)

TSXV: ELE   OTCQX: ELEMF

elementalaltus.com

Market capitalization: $79 million

Elemental Royalties Corporation began public trading in 2020. The company changed its name in 2022 after the merger of Elemental and Altus Strategies.[6] Boasting royalties across the Americas, Africa, and Australia, the company uses its quality acquisition model to seek and acquire advanced royalties in both producing and pre-producing mines. Elemental funds its projects and generates income using joint venture partnerships. Rather than acquiring a large number of exploration royalties, the company builds on the growth

---

5. "Our Story," Electric Royalties, 2022, www.electricroyalties.com.

6. "Elemental Royalties Announces Name Change to Elemental Altus Royalties Corp.," Elemental Royalties, September 22, 2022, www.elementalaltus.com.

value of revenue-generating projects and prides itself on acquiring clean royalties without caps or buybacks.[7]

EMPRESS ROYALTY CORPORATION

TSXV: EMPR   OTCQB: EMPYF

empressroyalty.com

Market capitalization: $25 million

We believe the royalty and streaming space will significantly evolve in the next 5 to 10 years, especially for the junior and mid-sized companies. The first shift for the junior industry will be to focus on mining companies and their project financing needs in order to get mines into production, that is, creating new royalties and streams as opposed to trading third-party royalties or purchasing portfolios. This shift of focus will result in royalty and stream financing structures being equally considered as a viable financing option for junior mining companies alongside traditional equity and debt. The second evolution will be further and more aggressive consolidation of junior and mid-sized royalty companies. The third will be royalty and streaming companies expanding their offerings to provide exclusive and unique full financing solutions (equity plus debt plus royalty or stream) to compete with the private equity competition.

New royalty and stream structures are used for the purpose of financing mines to get into production or expand on current production capacity; the needed capital goes directly to the mining company. Alternatively, buying existing royalties or a royalty portfolio does not help mining companies raise the funding they need, instead the capital goes to a third party, whom often has no relation to the mining company. In addition to not financing a mining project, every time a third-party royalty is traded by a new group, the premium increases and is paid by the purchaser. There are a finite amount of existing third-party royalties on mining projects at the moment, and they

---

7. "Connecting Investors to Gold," Elemental Altus Royalties Corporation, 2022, www.elementalaltus.com.

are being acquired and/or traded faster than new ones are being created. This results in competitive bidding processes, paying higher premiums, and no benefit to mining companies.

By creating new royalty and stream structures, Empress Royalty Corporation is able to offer a new financing option to the mining company that can be provided alone or with a combination of debt/equity. We are able to target our capital into near-term or immediate revenue-generating investments, which create significant value for both Empress shareholders and our mining company partners. The creation of new royalties and streams result in better internal rates of returns, close relationships with the mining companies (that de-risk the investment), and in-depth knowledge of their future plans leading to potential upside and further investment opportunities.

In summary, we believe that over the next 5 to 10 years, the junior royalty space will see three major changes:

1. The focus will shift from low-return third-party royalty trading or acquisitions to our business model of financing mining companies through high-return royalty and stream creation.
2. There will be further consolidation of junior royalty companies and the senior stream companies.
3. Streaming and royalty companies will expand their offering to provide exclusive and unique full financing solutions to compete with the private equity competition.

EMX ROYALTY CORPORATION

NYSE | TSXV: EMX   FSE: 6E9

emxroyalty.com

Market capitalization: $231 million

EMX Royalty Corporation has a long-standing track record of success in exploration discovery, royalty generation, royalty acquisition, and strategic investments. Our diversified, three-pronged business approach provides exposure to multiple upside opportunities, while minimizing the impact on EMX's treasury.

EMX's business model is designed to efficiently manage the risks inherent to the minerals exploration and mining industry. Key elements and resulting advantages of our unique approach are as follows:

- We organically generate royalties through low-cost property acquisition and early stage exploration to build value and then develop partnerships with quality companies to advance the projects, with EMX retaining a royalty interest and receiving pre-production payments.
- Our organic royalty growth is supplemented by purchases of royalties from other parties, as well as strategic investments.
- Cash flow from royalties, advance royalties, and other property payments are supplemented by returns from strategic investments and provide self-funding operating capital for our ongoing business initiatives.
- Using this model, we sustainably grow the royalty portfolio, with minimal dilution to our shareholders.

EMX's royalty and property portfolio spans five continents, and consists of a balanced mix of precious metal, base metal, and other assets.[8] In 5 or 10 years, we suspect that there will be more mid to large royalty companies via consolidation. As the royalty arena continues to become more competitive and available royalties scarcer and more capital intensive to acquire, the most expeditious avenue for growing a royalty company will be mergers or acquisitions. We have seen a similar phenomenon in the mining sector, and we expect the royalty sector to follow suit.[9]

Another means for growing a royalty portfolio in a very cost-efficient manner is *royalty generation*; a concept which EMX Royalty conceived, refined, and

8. "Royalty Cash Flow, Generation, Acquisition; Strategic Investment," EMX Royalty Corporation, 2022, https://emxroyalty.com.

9. David Cole to Douglas Silver, personal communication, April 14, 2022.

actively applies. We couple this with conventional royalty acquisition and strategic investment.

EURO RESSOURCES
NYSE Euronext Paris: EUR

goldroyalties.com

Market capitalization: $201 million

EURO Ressources is a precious metals royalty company. IAMGOLD France, which is an indirect wholly owned subsidiary of IAMGOLD Corporation, owns most of the stock shares and voting rights of EURO Ressources. In addition to marketable securities, the main assets of EURO Ressources are royalties on Suriname's Rosebel gold mine production, the Paul Isnard exploration project in French Guiana, and a silver stream in Burkina Faso. Rosebel is owned and operated by IAMGOLD. Paul Isnard is a joint venture project and is a net smelter return (NSR) production royalty.

GREAT BEAR ROYALTIES CORPORATION[10]
TSXV: GBRR

greatbearroyalties.com

Market capitalization: $121 million

Great Bear Royalties Corporation was formerly a Toronto Stock Exchange Venture Exchange (TSXV)-listed precious metals royalty and streaming company. The company was acquired by subsidiaries of Royal Gold (together called *Royal Gold*) on September 9, 2022 for cash consideration of Can$6.65 per common share for aggregate consideration of approximately Can$199.5

---

10. All stock exchange and market capitalization information for Great Bear Royalties Corporation is as of December 31, 2021.

million. Following the acquisition, the company was delisted from the TSXV.[11]

The company's principal asset was a 2% NSR royalty on Kinross Gold Corporation's (TSX: K) Great Bear project located in northwestern Ontario (Canada). In exchange for certain information and access provided to Royal Gold by Kinross in connection with Royal Gold's acquisition of the company, Kinross acquired the option to purchase a 25% interest in the royalty (0.5% of the 2.0% royalty rate) for an amount equal to 25% of Royal Gold's purchase price, adjusted for inflation, at any time until the earlier of a construction decision for the project and 10 years after the transaction closing date. The Great Bear project was acquired by Kinross from Great Bear Resources (TSXV: GBR) for $1.4 billion on February 24, 2022.[12]

LABRADOR IRON ORE ROYALTY CORPORATION
TSX: LIF

labradorironore.com

Market capitalization: $1,877 million

Labrador Iron Ore Royalty Corporation (LIORC) is a publicly traded Canadian company. LIORC holds interest in the Iron Ore Company of Canada (IOC), which has been mining, processing, and exporting iron ore concentrate and pellets since 1954. LIORC also leases land to IOC, and IOC pays LIORC a royalty on all the sales of all the products mined on the LIORC land.[13]

---

11. Royal Gold, "Royal Gold Completes Acquisition of Great Bear Royalties Corp." press release, September 9, 2022, www.marketscreener.com.

12. "Corporate," Great Bear Royalties, 2022, www.greatbearroyalties.com.

13. "Company Profile," Labrador Iron Ore Royalty Corporation, 2022, https://labradorironore.com.

MESABI TRUST

NYSE: MSB

mesabi-trust.com

Market capitalization: $347 million

Mesabi Trust, a mining royalty trust, generates income from the Peter Mitchell taconite mine (Minnesota, United States). Northshore Mining Company, which is a subsidiary of Cleveland-Cliffs (CCI) operates the mine. The taconite is processed into iron ore pellets, which are used in steelmaking. CCI is the largest producer of iron ore pellets in North America.[14]

MORIEN RESOURCES CORPORATION

TSXV: MOX   OTC: APMCF

morienres.com

Market capitalization: $8 million

Morien Resources Corporation is a Canada-based, mining-development company that holds royalty interests in two tidewater-accessed projects. Before going on care and maintenance, the Donkin coal mine (Nova Scotia, Canada) commenced production in 2017, and the Black Point aggregate project (Nova Scotia, Canada), permitted in 2016, is progressing toward a development decision and is paying advanced minimum royalties to Morien. The company's management and its board of directors consider shareholder returns to be paramount over corporate size, number or scale of assets, and industry recognition.[15]

---

14. "About Mesabi Trust," Mesabi Trust, 2008–2022, www.mesabi-trust.com.

15. "About Morien Resources," Morien Resources Corporation, 2022, https://morienres.com.

MUNDORO CAPITAL
TSXV: MUN   OTCQB: MUNMF

mundoro.com

Market capitalization: $17 million

Mundoro Capital is a Canadian-listed royalty-generator company with a portfolio of projects focused on primarily base metals that generate royalties and near-term revenue, that is, generate cash from the portfolio. To drive value for shareholders, Mundoro has assembled a portfolio of mineral projects chiefly focused on copper and gold in two mineral belts: the Western Tethyan belt in eastern Europe and the Laramide belt in southwestern United States. Value generated from our mineral properties is through near-term revenues from various forms of payments from partners and long term through the generated royalties.

Royalty Sector Outlook

The outlook on the sector depends on various factors. If we concentrate on two possible scenarios in the gold market, we can broadly anticipate what the effect could be on the royalty sector for the traditional precious metal–focused royalty companies:

1. Higher gold environment. If gold breaks $3,000 per ounce, then we think we will continue to see the marketplace play out as it already has over the last five years, which is additional new royalty entrants and heavier competition for existing royalties on producing and development mines that will result in significant increase in valuations for royalties broadly. Based on some valuations for precious metal royalties in the marketplace, a fair portion of royalty companies are likely taking this view as to when acquisition prices are taken into consideration.

2. Lower, sideways gold environment. If gold does not break out of its current trading range of around $2,000 per ounce plus or minus 20%, then we think there will be (1) a consolidation of royalty companies on the precious metal side and (2) a proliferation of royalty companies seeking base metal royalties (with the immediate focus on the battery

metals). However, there are a significantly fewer number of base metal operations and assets for which to compete for royalties, which will cause valuations of base metal royalties to run up quickly. The base metal mines are more tightly concentrated among a small group of producers. This will cause fragmentation of royalties into smaller pieces and create an active secondary market. The obvious and larger market is alternative energy royalties. We have seen a smaller subset of royalty companies venture into this market segment in prior market cycles, and if we assume the gold price does not enter a higher gold environment, then this diversification into the energy royalties will be an important beachhead to start transitioning into now. If the institutional investor base is investing for the royalty structure of stable cash flow and dividends over the long term, then this should be a welcome diversification strategy into new long-life revenue streams; however, if the investor base is interested in the precious metals exposure, then royalty companies following this strategy will likely be penalized.

Diversification is a differentiating factor for growth in the metals royalty market. To build attention with a broader institutional investor base (beyond the traditional gold-focused funds), large-cap royalty companies will need to become less dependent on gold royalties and broaden their portfolios to diversify their risk profile over commodity cycles as their shareholder base becomes increasingly dividend driven such as the institutional pension-related asset managers. Although the gold premium will continue to exist to a certain extent, it will likely become outweighed by the need for stable cash flow and dividends across commodity cycles for the institutional asset managers. Mid-cap royalty companies will likely follow the path of the large-cap royalty companies and diversify their asset base. To differentiate themselves, they may venture into parallel royalty market segments sooner than the senior royalty companies.

Royalty Sector Strategy

Perhaps the most interesting segment of the royalty market is the small-cap exploration royalty companies that generate new royalties. This is where Mundoro is active. To explain exploration in the context of the broader market, exploration is the research portion of an industry. Recognizing that research broadly has a success rate of 1%, this translates into the minerals sector that 1% of targets generated in exploration succeed to a mineral discovery and of those, about 10% advance to economic development, and of those, approximately 30% advance to future production. As a result, the small-cap exploration royalty companies that focus on generating a portfolio of royalties on exploration assets are creating new royalties at a fraction of a penny on the dollar of the cost to purchasing a royalty of one producing mine.

Because of the perceived risk of valuing exploration, succeeding in the exploration royalty segment takes a different type of valuation model and mindset that we do not see the large-cap and mid-cap royalty companies being willing to transition into, which leaves a valuable opportunity for Mundoro. Because we generate our properties and then bring in international senior base metal miners as our partners to whom we option out those properties, we are in the strategic position to establish the valuation for the future royalty and the annual payment stream. Mundoro collects annual property payments from time of the deal signing, receives milestone payments along the exploration and development timeline as the properties advance, and retains a future NSR royalty over production. We have essentially widened the concept of a paying royalty upstream right into the early exploration phase.

Why Investors Should Pay Attention to Exploration Royalty Companies

Mundoro is an opportunity to own a company that is already generating annual payments from its portfolio, which consists of multiple exploration base metal–focused assets, on properties optioned out to senior mining producers (i.e., the counterparties) Thus, Mundoro provides a discovery upside from the nature of the underlying asset. The counterparties Mundoro transacts with are those that the large-cap royalty companies are seeking to hold royalties with. Hence, Mundoro is not only building a valuable model of generating annual cash flow from exploration properties, but Mundoro could also be considered a valuable takeover candidate for large-cap royalty companies seeking exposure to base metal royalties on assets owned by the senior mining producers.[16]

NICKEL 28 CAPITAL CORPORATION

TSXV: NKL   FSE: 3JC0

nickel28.com

Market capitalization: $69 million

Nickel 28 Capital Corporation is a nickel and cobalt–focused investment vehicle with an emphasis on metal streaming and royalty agreements, offering investors exposure to metals integral to key technologies of the electric vehicle and energy storage markets.[17] The company is one of the only operating, cash flow–positive nickel-specific vehicles listed in Canada. Thus, the company intentionally offers investors exposure to nickel price moves.

Nickel 28 was formed as a spin-off when Cobalt 27 was taken over. The acquirers of Cobalt 27 bought the cobalt stream on Voisey's Bay, and the physical cobalt and all of the nickel assets in the Cobalt 27 portfolio were spun out as Nickel 28.

Nickel 28's key operating asset is a joint venture interest in the Ramu nickel mine in Papua New Guinea. The company's partner is the Metallurgical Corporation of China (MCC). The mine is the largest producer of MHP, which is a mixed nickel-cobalt hydroxide product, in the world and has been operating for nearly 8 years with an anticipated mine life of over 40 years. The company's royalty portfolio is a range of projects covering North America and Australia, with the royalties on the Dumont (Québec) and Turnagin (British Columbia) projects in Canada most likely to come in to production first.

---

16. Teo Dechev to Douglas Silver, personal communication, July 27, 2022.

17. "About Us: Company Profile," Nickel 28 Capital Corporation, 2022, www.nickel28.com.

The success of investment vehicles, such as Franco-Nevada Corporation and Wheaton Precious Metals Corp., inspired Anthony Milewski to found Cobalt 27. The original idea behind Cobalt 27 was to take a stockpile of physical cobalt and use it as a balance sheet to acquire cobalt streams. The creation of Cobalt 27 coincided with an explosion in interest in electric vehicles and the cultural phenomenon of Elon Musk pushing the adoption of electric vehicles. Cobalt 27's good fortune in timing allowed it to raise more than a billion dollars in equity and debt in two short years. At one point, the market capitalization of the company was nearly $1 billion. The cobalt price cycle, however, was shorter than expected, and the forward-looking demand could not support a commodity price move from $8 to $44 in such a short period. The price of the commodity collapsed, causing the stock price to follow and ultimately open the door for an investor to accumulate a block of stock and take the company over. The management team tried to expand the Cobalt 27 portfolio past cobalt to include nickel but was too late to do so and keep the assets together. Switzerland-based Pala Investments was ultimately successful in acquiring an outsized equity position.

There are very few "streaming and royalty" investors. Instead, the large market cap royalty companies have a shareholder base that is focused on precious metals and prefer to invest in the streaming and royalty structure. This means that the valuation metrics that non-precious metal streamers often try to attribute to their stock as a way demonstrating they are undervalued is flawed as it presupposes the same investor pool. It also limits the capital-raising opportunities of non-precious metal juniors to bull markets for their given commodities.

Hyper-focused small streaming and royalty companies are highly susceptible to commodity moves. These commodity moves impact their share price and cost of capital. In a down cycle, the cost of capital at a small streaming and royalty company means that shareholders have to suffer what might be significant dilution in order to add new assets to the portfolio.

It may turn out that the junior streaming and royalty companies act as aggregators of assets for the senior royalty companies to roll up. Private equity firms, such as Orion Resource Partners (mine financing), have successfully created junior companies that were later acquired. That said, the entire sector from the largest companies to the smallest has shown minimal interest in merging because of an employee-light management model that can go on for years with little to no capital injections. Without a focused shareholder base, there are few catalysts for management to want to give up their seats and merge.

NOMAD ROYALTY COMPANY

TSX | NYSE: NSR

nomadroyalty.com

Market capitalization: $429 million

Nomad Royalty Company was a gold and silver royalty company that purchased rights to a percentage of the gold or silver produced from a mine, for the life of the mine. Nomad owned a portfolio of 20 royalty and stream assets, of which 8 were on producing mines. Nomad planned to grow and diversify its low-cost production profile through the acquisition of additional producing and near-term producing gold and silver streams and royalties.[18]

Postscript

On May 2, 2022, Nomad Royalty Company agreed to be acquired by Sandstorm Gold.[19] In an all-share transaction valued at approximately Can$755 million. Nomad shareholders received 1.21 Sandstorm common shares for each Nomad share held. The consideration implies a value of approximately Can$11.57 per Nomad share based on the closing price of Sandstorm's shares on the Toronto Stock Exchange (TSX) on April 29, 2022 and represents a 21% premium to closing price of Nomad shares on

---

18. Nomad Royalty Company, "Nomad Royalty Reports Q1 Results and Declares Second Quarter 2022 Dividend," press release, May 5, 2022.

19. Nomad Royalty Company, "Nomad Royalty Announces Friendly Acquisition by Sandstorm Gold," press release, May 5, 2022.

the same date and a 34% premium on the 20-day volume-weighted average price (VWAP).[20]

NOVA ROYALTY CORPORATION

TSXV: NOVR   OTCQB: NOVRF

novaroyalty.com
Market capitalization: $192 million

Nova Royalty Corporation was founded in 2018 as a first mover in the copper royalty space. The company sees copper as the world's most strategic commodity, which is the key element in the clean energy transition, including electric vehicles. In addition, the company also acquires royalties on strategic nickel deposits, which can drive the expansion of developed-market electric vehicle supply chains. Nickel is the key metal ingredient in electric vehicle batteries, accounting for more than 70% of the metal content in dominant battery chemistries.

The company has assembled a diversified portfolio of royalties on major copper and nickel projects already being advanced by leading mining companies, generally cost more than $3 billion to build, and once constructed, often last for more than 50 years of operations. This portfolio gives Nova a unique and irreplicable foundation to become the lowest cost-of-capital player in its commodity space and sets the scene for Nova to become a senior, Tier 1 royalty company.

Since inception, the company has deployed approximately $65 million of capital in acquiring royalties on the most significant emerging projects in copper and nickel that are being advanced by major mining companies, including the Taca Taca project in Argentina, owned by First Quantum Minerals; Copper World Complex in Arizona (United States), owned by Hudbay Minerals; Josemaria project in Argentina, owned by Lundin Mining; West Wall project in Chile, owned by AngloAmerican and Glencore; and the NuevaUnion

20. Nomad Royalty Company, "Nomad Royalty Reports Q1 Results and Declares Second Quarter 2022 Dividend," press release, May 5, 2022.

project in Chile, owned by Teck Resources and Newmont Corporation. Nova also owns royalties on the Vizcachitas copper project in Chile and the Dumont nickel project in Québec (Canada), both of which it expects to transition to major strategic partners in the next several years. The company is also planning a U.S. listing after completing additional acquisitions.

Nova's cash flows, free from any capital or operating cost obligations, give it a highly attractive profile in comparison to producers, trading houses, smelters, or any other entities on the supply chain of the copper and nickel markets, which already have total annual sales of over $300 billion and are poised for transformational growth in the years ahead.

Nova's foundation of royalties would be impossible to replicate for any new entrant. The number of quality camps is limited, thus giving the first mover a critical advantage. Further, the royalties that Nova has acquired were sourced by its global team of more than 20 people located across all of the major mining camps in the Americas and Australia. The local sourcing model relies on acquiring existing royalties on top-tier deposits from the original prospectors who sold the assets to the major companies 30 to 40 years ago in exchange for a royalty. That is the only known way to acquire a royalty on a top-tier project as none of the major mining companies have ever sold a royalty on the primary commodity of their core assets.[21]

OROGEN ROYALTIES

TSXV: OGN   OTCQX: OGNRF

orogenroyalties.com

Market capitalization: $65 million

Orogen Royalties holds gold and copper royalties in projects throughout North America, but considers the Ermitaño gold deposit (Sonora, Mexico) and Silicon gold project (Nevada, United States) as the company's two most important assets. Both royalties are based on NSR structures. Using a business

---

21. "Nova Royalty Corporation (TSX Venture: NOVR)," Nova Royalty Corporation, 2022, www.novaroyalty.com.

model of prospect generation, Orogen prides itself on identifying and moving forward high-quality projects using its geological and financial expertise.[22]

SAILFISH ROYALTY CORPORATION

TSXV: FISH   OTCQX: SROYF

sailfishroyalty.com
Market capitalization: $87 million

Sailfish Royalty Corporation is a precious metals royalty and streaming company. Within Sailfish's portfolio are two main assets in the Americas: a gold stream equivalent to a 3% NSR on the San Albino gold mine (approximately 3.5 square kilometers) and a 2% NSR on the rest of the area (about 134.5 square kilometers) surrounding San Albino in northern Nicaragua; and an up to 3% NSR on the multimillion-ounce Spring Valley gold project in Pershing County, Nevada (United States).[23]

SCULLY ROYALTY

NYSE: SRL

www.scullyroyalty.com
Market capitalization: $131 million

Scully Royalty is a worldwide business, incorporated in the Cayman Islands. The company includes an iron ore royalty (Newfoundland and Labrador) in Canada, industrial equity focused on eastern Asia, and a holding company listed on the Malta stock exchange. Scully claims the mine produces a premium-quality iron ore product with high iron (Fe) content. Thus the company focuses on providing high earnings and dividends for its shareholders from the iron ore royalty.[24]

---

22. "About Orogen," Orogen Royalties, 2022, www.orogenroyalties.com.

23. "Streams & Royalties," Sailfish Royalty Corporation, 2021, https://sailfishroyalty.com.

24. "About Scully," Scully Royalty, accessed November 1, 2022, www.scullyroyalty.com.

STAR ROYALTIES

TSXV: STRR   OTCQX: STRFF

starroyalties.com

Market capitalization: $35 million

Star Royalties is a precious metals and green royalty and streaming investment company. The company created the world's first carbon-negative gold royalty platform and offers investors gold exposure with an increasingly negative carbon footprint. The company's objective is to provide wealth creation through accretive transaction structuring and asset life extension with superior alignment to both counterparties and shareholders.[25]

TRIDENT ROYALTIES

AIM: TRR   FSE: 5KV

tridentroyalties.com

Market capitalization: $122 million

Trident Royalties is a diversified mining royalty and streaming company, providing investors with exposure to base and precious metals, bulk materials (excluding thermal coal), and battery metals.[26] Since inception, Trident has proven itself to be one of the fastest growing royalty companies, having listed on the alternative investment market (AIM) of the London Stock Exchange with a single royalty and, in less than two years, grown to a portfolio of more than 20 assets, with over half being cash flow generative. Trident's portfolio presently includes exposure to iron ore, copper, gold, and lithium; however, it is the intent of the company to broaden its commodity diversification such that a shareholding in Trident will provide investors with commodity exposure covering the full breadth of the mining sector. As such, over the coming years, Trident expects to continue to diversify its portfolio in an accretive

25. "Royalties to Enable a Carbon Negative Future," Star Royalties, 2022, www.starroyalties.com.

26. "Strategy," Trident Royalties, accessed October 11, 2022, https://tridentroyalties.com.

manner, ultimately providing exposure to a broader sweep of mining commodities, inclusive of precious metals; base metals; and bulks, industrials, and battery minerals. In tandem with this targeted growth, key assets currently within Trident's portfolio are expected to move into production, providing a significant foundation of cash flow from which to further grow the company, while reducing the cost of capital and providing line-of-sight to an inaugural dividend.[27]

URANIUM ROYALTY CORPORATION

TSXV: URC   NASDAQ: UROY

uraniumroyalty.com

Market capitalization: $390 million

Uranium Royalty Corporation (URC) was formed in 2018 as the first pure-play uranium royalty company in the world. It launched publicly in December 2019 through a Can$30 million IPO on the TSXV (the largest mining IPO in Canada for 2019). The IPO was led by founder, Amir Adnani (chairman) and Scott Melbye (CEO). Adnani and Melbye had recognized that despite the robust royalty and streaming industry, which had emerged in the base and precious metals industries, no such vehicle existed in the uranium space.

URC attracted the early backing of notable resource financiers Marin Katusa and Warren Gilman, along with resource funds such as Extract Capital and Sprott Asset Management. Additionally, two corporates—Uranium Energy Corporation (UEC) and Mega Uranium—vended in royalties on projects based in the United States and Namibia. UEC and Mega both saw the inherent value in realizing a re-rating on royalties they held, which were then trading at a big discount to net asset value at a time where precious metal royalty companies were trading at large premiums to net asset value.

---

27. Adam Davidson to Douglas Silver, personal communication, July 27, 2022.

Early on, URC recognized that there was significant upside present in the uranium markets with pricing below both incentive price levels and the operating costs of existing mines. This outlook resulted in URC taking a unique track compared to its precious metals peers and focusing on gaining exposure to physical uranium. As a result, URC made a major pre-IPO acquisition, committing $20 million as a foundational 9.9% investor in the IPO of Yellow Cake (YCA) in London. Not only did this investment effectively provide URC investors exposure to uranium near the cycle lows (at approximately $21 per pound), but it also offered the benefits of a strategic relationship with YCA. The most significant was the right to access over $31 million worth of physical uranium under a $1.07 billion purchase agreement with the world's largest uranium producer, Kazatomprom, the national operator for Kazakhstan.

The first such uranium purchase under that option occurred in early 2021 when 348,068 pounds uranium oxide (U_3O_8) were secured at $28.73 per pound. Subsequent market purchases in 2021 and early 2022 resulted in a physical uranium holding of 1.55 million pounds at an average cost of $42.20 per pound. The rationale behind this successful physical uranium strategy was to capture attractively priced uranium as a liquid and appreciating asset.

Another key development for URC was the 2021 acquisition of royalties on the world's two largest and highest-grade uranium mines, Cigar Lake and McArthur River in Saskatchewan, Canada. URC acquired these royalties, not from the majority owners, Cameco Corporation and Orano Group, but rather from Frank Melfi of Reserve Minerals in Albuquerque, New Mexico. Melfi, now in his mid-90s, was instrumental in the exploration activities in the late 1970s and early 1980s when these word-class deposits were discovered through private joint ventures. When ownership of these deposits was rolled up by provincial and federal crown corporations at the time, it was typical for the successful explorers to retain a royalty interest.

Forming URC near decade lows in the uranium price was a contrarian decision. However, Adnani and Melbye had a strong conviction that nuclear energy would again enjoy broad-based support through its ability to provide safe, carbon-free, and resilient 24/7 baseload power to an energy-hungry world. This global growth in nuclear power will require substantial new investment in the next generation of mines to produce the needed uranium fuel, and URC is well positioned to provide that development funding through royalties and streams.

URC was formed in 2018 with uranium price at $20 per pound; in April 2022, the uranium price is at $60 per pound, driven by a global movement to decarbonize the economy while scaling electrification.

URC has royalty assets in all of North America's major global uranium districts, whether that be emerging in situ recovery mines in the United States, high-grade mines in Saskatchewan, and an interest in a major, proven African producer. URC is a $500 million market capitalization company in 2022 with a strong platform to grow into a $5 billion company to the end of this decade with the backdrop of unprecedented growth in demand for uranium and structural supply deficit from mining.[28]

VOX ROYALTY CORPORATION
TSXV: VOX
voxroyalty.com
Market capitalization: $107 million

Vox Royalty Corporation is a junior royalty company focused on obtaining or creating precious metals royalties. The company was formed in 2014 and has a current portfolio of more than 50 royalties and streams spanning eight jurisdictions.

---

28. "Company: About," Uranium Royalty Corporation, 2022, www.uraniumroyalty.com.

VOYAGEUR MINERAL EXPLORERS CORPORATION
CSE: VOY

voyageurexplorers.com
Market capitalization: $12 million

Voyageur Mineral Explorers Corporation is a Canadian junior mineral exploration company with a specific focus on mineral properties in northwest Manitoba and northeast Saskatchewan, Canada. The company owns a valuable package of royalties in the prolific Flin Flon greenstone belt and has assembled a portfolio of base metal and precious metal prospects in Manitoba and Saskatchewan.[29]

XTIERRA
TSXV: XAG

xtierra.ca
Market capitalization: $1 million

Xtierra provides direct exposure to mining revenues from the recent growth in precious metal prices. Our goal is to increase share value by accumulating a diversified portfolio of cash-flowing royalties to pay increasing dividends.[30]

---

29. "Home," Voyager Mineral Explorers Corporation, 2022, https:// voyageurexplorers.com.

30. "Focused on Cash Flowing Gold, Silver, and Copper Royalties," Xtierra, 2020, www.xtierra.ca.

METHODOLOGIES

Doug Silver

This study was complicated because it involved 54 companies and 2,000 assets over a 36-year timeline, so a systematic methodology had to be established to organize and keep the data consistent. The data cutoff date was December 31, 2021, to allow time for data compilation, analysis, and synthesis.

Because of the availability of data, the study only considers public mining royalty companies. Therefore, if a company was private and subsequently went public, its financials were tracked from the date of going public. Similarly, several companies began as exploration or mining companies and subsequently converted to the royalty and/or streaming format. Efforts were made to identify this conversion date by reading through press releases or contacting the company directly. The financials were then tracked from that point forward.

Many of these companies were acquired by other companies. It is assumed that the financials for the targeted company were assimilated into and reflected in the predator company's financials in the year in which it was taken over.

Initially, a list of active royalty companies was compiled and later augmented with additional royalty companies that had been purchased by these companies. Each company's website was then reviewed for its relevant information. This was easy for the larger companies because most publish asset handbooks. Asset-specific data was extracted from websites, annual reports, annual information forms, press releases, and investor relations presentations. Additional data was also pulled from the project operator's website, reports, and press releases. Finally, companies were contacted for missing information.

It should be noted that SEDAR[1] only contains files from 1997 forward and EDGAR[2] began in 2001. I particularly thank Franco-Nevada Corporation for digging out old annual reports for four companies that they built or purchased.

Market capitalizations were calculated using the annual closing share price as listed in Stockwatch.[3] Shares outstanding at year's end were taken from annual reports or by averaging the shares outstanding from press releases that straddle the end of the year. No efforts were made to include in-the-money options in these calculations.

All currencies were converted to U.S. dollars using monthly rates as reported by the Pacific Exchange Rate Service.[4] Not all companies have a December fiscal year end, but this study is investigating industry trends and the quarter-to-quarter variability is not extreme.

A database of royalties, offtakes, and streams was created. This often includes more assets than are currently reported on the companies' website for several reasons. Companies will lump together or average different royalties if they exist on the same claims or property. Therefore, the exact number of royalties is unknown.

Offtakes and streams are more discrete. Based on available data, the known assets were unbundled when possible. For example, if a company had two royalties on the same project, but each had a different rate, the royalty was then disaggregated into two royalties. Similarly, if different metals on a given project each had a different royalty rate, these were also separated. If a company

---

1. System for Electronic Document Analysis and Retrieval—The Canadian Securities Administrators' document depository.

2. Electronic Data Gathering, Analysis and Retrieval System—The U.S. Securities and Exchange Commission's document depository.

3. "Quotes, Charts, News, Indexes, Portfolio, Analytics, Mutual Funds," www.stockwatch.com.

4. Pacific Exchange Rate Service, University of British Columbia, Sauder School of Business, https://fx.sauder.ubc.ca.

owned a royalty on a given property and subsequently acquired additional royalty interests, these were considered as separate royalties.

Situations exist where a royalty company obtains a portfolio of royalties in a similar district but with noncontiguous land positions. In this situation, each isolated land package was considered as a separate asset. The study also included royalties that had expired prior to 2021.

The analysis for this book is based on best available data. Despite extensive efforts to capture all of the data, this was not possible. Specifically, I was unable to obtain annual reports for Mesabi Trust (1986–1992), Anglo Pacific Group (2001–2006), Repadre Capital (1992–1994), and Redstone Resources (1995).

The larger companies focus on their producing and advanced-stage assets and tend to table their early stage exploration royalties with less disclosure. It should also be noted (without criticism) that British and Australian companies release a lot less information than U.S. and Canadian companies, so tracking their project specifics can be very difficult, if not impossible.

Today's companies are so large that some are taking equity positions in other companies, purchasing oil and gas working interests or other non-mineral investments such as power plants. These assets are excluded from this study but included when presenting corporate financial statements.

Despite these caveats, I was able to collect more than 23,000 pieces of corporate information and approximately 100,000 pieces of asset information. This should be sufficient for validating the industry trends.

INDEX

Note: *f.* indicates figure; *t.* indicates table